QUANTUM PHYSICS VOYAGE

BEGINNERS GUIDE FROM STRING THEORY TO QUANTUM COMPUTING

4 BOOKS IN 1

BOOK 1
QUANTUM PHYSICS FOR BEGINNERS: EXPLORING THE FUNDAMENTALS OF QUANTUM MECHANICS

BOOK 2
FROM STRING THEORY TO QUANTUM COMPUTING: A JOURNEY THROUGH QUANTUM PHYSICS

BOOK 3
QUANTUM PHYSICS DEMYSTIFIED: FROM NOVICE TO QUANTUM EXPERT

BOOK 4
MASTERING QUANTUM PHYSICS: FROM BASICS TO ADVANCED CONCEPTS

ROB BOTWRIGHT

Published by Rob Botwright
Library of Congress Cataloging-in-Publication Data
ISBN 978-1-83938-623-7
Cover design by Rizzo

Disclaimer

The contents of this book are based on extensive research and the best available historical sources. However, the author and publisher make no claims, promises, or guarantees about the accuracy, completeness, or adequacy of the information contained herein. The information in this book is provided on an "as is" basis, and the author and publisher disclaim any and all liability for any errors, omissions, or inaccuracies in the information or for any actions taken in reliance on such information. The opinions and views expressed in this book are those of the author and do not necessarily reflect the official policy or position of any organization or individual mentioned in this book. Any reference to specific people, places, or events is intended only to provide historical context and is not intended to defame or malign any group, individual, or entity. The information in this book is intended for educational and entertainment purposes only. It is not intended to be a substitute for professional advice or judgment. Readers are encouraged to conduct their own research and to seek professional advice where appropriate. Every effort has been made to obtain necessary permissions and acknowledgments for all images and other copyrighted material used in this book. Any errors or omissions in this regard are unintentional, and the author and publisher will correct them in future editions.

BOOK 1 - QUANTUM PHYSICS FOR BEGINNERS: EXPLORING THE FUNDAMENTALS OF QUANTUM MECHANICS

BOOK 2 - FROM STRING THEORY TO QUANTUM COMPUTING: A JOURNEY THROUGH QUANTUM PHYSICS

BOOK 3 - QUANTUM PHYSICS DEMYSTIFIED: FROM NOVICE TO QUANTUM EXPERT

BOOK 4 - MASTERING QUANTUM PHYSICS: FROM BASICS TO ADVANCED CONCEPTS

Introduction

Welcome to the "Quantum Physics Voyage," a comprehensive bundle of books that will take you on an extraordinary journey through the captivating world of quantum physics. Whether you are a curious beginner or an aspiring quantum expert, this collection of books is designed to be your trusted companion as you embark on an exploration of the fundamental principles and cutting-edge concepts that define the quantum universe.

Book 1, "Quantum Physics for Beginners: Exploring the Fundamentals of Quantum Mechanics," serves as your passport to the quantum realm. In these pages, we will unravel the mysteries of quantum mechanics, diving deep into the fundamental concepts that govern the behavior of the smallest particles in the universe. From wave-particle duality to the intriguing phenomena of quantum superposition and the uncertainty principle, this book will provide you with a solid foundation to understand the quantum world.

Book 2, "From String Theory to Quantum Computing: A Journey Through Quantum Physics," invites you to embark on an exhilarating voyage that spans from the elegant principles of string theory to the revolutionary field of quantum computing. We will explore the unification of quantum mechanics and general relativity, opening the doors to a new understanding of the cosmos. Additionally, we will delve into the world of qubits, quantum algorithms, and the promise of quantum supremacy in computation.

Book 3, "Quantum Physics Demystified: From Novice to Quantum Expert," marks the next phase of our voyage. Here, we transition from novice to quantum enthusiast by tackling advanced topics in quantum mechanics. Quantum states, operators, and experiments will become familiar territory as we deepen our understanding of this fascinating field. This book serves as a bridge that empowers you to delve deeper into quantum physics.

Finally, in Book 4, "Mastering Quantum Physics: From Basics to Advanced Concepts," we ascend to the summit of our journey. Armed with a profound knowledge of quantum mechanics, we explore advanced concepts that include quantum field theory, relativistic quantum mechanics, and the intriguing domain of quantum gravity. Additionally, we examine the captivating connections between quantum physics and the enigmatic world of string theory.

The "Quantum Physics Voyage" bundle has been meticulously crafted to guide you through the intricate and awe-inspiring world of quantum physics. Each book builds upon the knowledge gained in the previous one, ensuring a seamless progression from fundamental principles to advanced concepts. Whether you are a student, a scientist, or simply a curious explorer of the universe, this bundle is your gateway to a deeper understanding of the quantum realm.

Prepare to embark on a journey that will challenge your perception of reality, spark your curiosity, and equip you with the knowledge to explore the mysteries of the quantum universe. Let us set sail on this extraordinary voyage, where the boundaries of classical and quantum physics blur, and the secrets of the cosmos await your discovery.

BOOK 1
QUANTUM PHYSICS FOR BEGINNERS
EXPLORING THE FUNDAMENTALS OF QUANTUM MECHANICS

ROB BOTWRIGHT

Chapter 1: The Quantum World Unveiled

The birth of quantum physics marked a monumental shift in our understanding of the fundamental nature of the universe. It emerged at the turn of the 20th century as a response to the limitations of classical physics, which had successfully described the behavior of macroscopic objects but struggled to explain the behavior of particles on the atomic and subatomic scale. As scientists delved deeper into the mysteries of the microscopic world, they encountered phenomena that defied classical intuition and demanded a new theoretical framework.

At the heart of this revolution were groundbreaking discoveries and the pioneering work of visionary physicists. One of the key figures in this development was Max Planck, who introduced the concept of quantization to explain the energy distribution of blackbody radiation. Planck's hypothesis that energy is quantized into discrete packets, or "quanta," marked the birth of quantum theory. It challenged the prevailing view that energy could be continuously divided and laid the foundation for a radical departure from classical physics.

Albert Einstein further advanced the field by applying the quantum concept to explain the photoelectric effect. In 1905, he proposed that light consists of discrete packets of energy called photons. This idea not only resolved the experimental observations of the photoelectric effect but also provided strong evidence for the existence of quanta in nature. Einstein's work on the photoelectric effect earned him the Nobel Prize in Physics in 1921 and solidified the acceptance of quantum theory.

The development of quantum theory gained further momentum with Niels Bohr's model of the hydrogen atom in 1913. Bohr's model incorporated the idea that electrons orbiting the nucleus could only occupy specific energy levels or "quantum states." This model successfully explained the spectral lines of hydrogen, which had previously baffled scientists. Bohr's quantization of electron orbits marked another pivotal moment in the quantum revolution.

However, it was Werner Heisenberg who introduced one of the most fundamental principles of quantum mechanics in 1925—the uncertainty principle. Heisenberg's principle stated that it is impossible to simultaneously measure certain pairs of properties, such as a particle's position and momentum, with absolute precision. This inherent uncertainty in quantum measurements challenged the deterministic worldview of classical physics and emphasized the probabilistic nature of quantum phenomena.

Meanwhile, Erwin Schrödinger formulated the Schrödinger equation in 1926, providing a mathematical framework to describe the behavior of quantum systems. This equation, often referred to as the cornerstone of quantum mechanics, allowed physicists to calculate the probability distribution of particles in various quantum states. Schrödinger's wave equation *was a major leap forward in our ability to understand and predict quantum behavior.*

Quantum mechanics also revealed the phenomenon of quantum entanglement, where two or more particles become intrinsically connected regardless of their spatial separation. Albert Einstein famously referred to this as "spooky action at a distance." The concept of entanglement challenged classical notions of locality and raised profound questions about the nature of reality.

Over the decades that followed, quantum physics continued to evolve and expand its reach. It provided the theoretical underpinnings for understanding the behavior of atoms, molecules, and the electromagnetic spectrum. Quantum mechanics played a pivotal role in the development of technologies such as lasers, transistors, and nuclear reactors.

In the mid-20th century, the advent of quantum field theory unified quantum mechanics with special relativity, resulting in a comprehensive framework to describe the behavior of particles and fields in the universe. This synthesis laid the groundwork for the Standard Model of particle physics, which successfully describes the fundamental particles and forces of the universe, except for gravity.

As quantum physics progressed, it also ventured into the realm of quantum computing and quantum information theory. The idea of using quantum bits, or qubits, as the fundamental unit of information introduced the possibility of exponentially faster computing and encryption-breaking algorithms.

In recent years, quantum physicists have explored the exciting field of quantum optics, where they manipulate individual photons and explore the fascinating phenomena of quantum teleportation and quantum cryptography. These advancements have the potential to revolutionize communication and computation in the future.

The birth of quantum physics not only reshaped our understanding of the physical world but also challenged our philosophical and metaphysical assumptions. It revealed a reality that is inherently probabilistic, where particles can exist in multiple states simultaneously and where the act of measurement itself influences the outcome. This departure from classical determinism sparked debates about the nature of free will, consciousness, and the ultimate nature of reality.

In summary, the birth of quantum physics marked a profound transformation in our understanding of the universe. From Planck's quantization to Schrödinger's wave equation, from Heisenberg's uncertainty principle to Einstein's photon hypothesis, the development of quantum theory reshaped the foundations of physics and opened the door to a new era of scientific exploration. Quantum physics has not only provided us with unprecedented technological advancements but has also challenged our worldview and invited us to ponder the profound mysteries of the quantum realm. It continues to be a frontier of discovery, offering glimpses into the deepest mysteries of the cosmos.

Quantum physics, a field that has reshaped our

understanding of the fundamental nature of reality, owes its existence to the pioneering work of brilliant scientists who dared to challenge the classical notions that had governed physics for centuries. The birth of quantum physics was marked by a series of key milestones, each representing a profound shift in our understanding of the physical world. One of the earliest pioneers in this journey was Max Planck, who, in 1900, introduced the concept of quantization to explain the energy distribution of blackbody radiation. Planck's revolutionary hypothesis that energy is quantized into discrete packets, or "quanta," was a fundamental departure from classical physics and laid the foundation for quantum theory. This concept not only explained the spectral distribution of energy but also led to the emergence of a new era in physics. Albert Einstein, another luminary of the era, further advanced quantum theory by applying it to the photoelectric effect in 1905. Einstein's groundbreaking proposal that light consists of discrete packets of energy called photons provided compelling evidence for the existence of quanta in nature. His work on the photoelectric effect was instrumental in solidifying the acceptance of quantum theory among the scientific community. Niels Bohr, a Danish physicist, made significant contributions to the field by introducing his model of the hydrogen atom in 1913. Bohr's model incorporated the idea that electrons orbiting the nucleus could only occupy specific energy levels or "quantum states," successfully explaining the spectral lines of hydrogen. This model was a major leap in understanding atomic structure and played a crucial role in the development of quantum physics. Werner Heisenberg, in 1925, introduced the famous uncertainty principle, a cornerstone of quantum mechanics. Heisenberg's principle stated that it is impossible to simultaneously measure certain pairs of properties, such as a particle's position and momentum, with absolute precision. This principle emphasized the inherent probabilistic nature of quantum measurements, challenging the deterministic worldview of classical physics. At the same time, Erwin Schrödinger formulated the Schrödinger equation, providing a mathematical framework to describe the behavior of quantum systems. This equation, often referred to as the cornerstone of quantum mechanics, allowed physicists to calculate the probability distribution of particles in various quantum states. Schrödinger's wave equation was a major step forward in our ability to understand and predict quantum behavior. These early milestones laid the foundation for the development of quantum mechanics as a comprehensive and revolutionary theory. Quantum mechanics introduced the concept of quantum states, which describe the properties and behaviors of particles, and operators, which represent physical observables and transformations. This framework allowed scientists to make precise predictions about the behavior of particles and to develop a deeper understanding of quantum phenomena. Quantum mechanics also revealed the phenomenon of quantum entanglement, where two or more particles become intrinsically connected regardless of their spatial separation. Albert Einstein famously referred to this as "spooky action at a distance," highlighting the profound and counterintuitive nature of entanglement. The concept of entanglement challenged classical notions of locality and raised profound questions about the nature of reality. Over the decades that followed, quantum physics continued to evolve and expand its reach. It provided the theoretical underpinnings for understanding the behavior of atoms, molecules, and the electromagnetic spectrum. Quantum mechanics played a pivotal role in the development of technologies such as lasers, transistors, and nuclear reactors. In the mid-20th century, the advent of quantum field theory unified quantum mechanics with special relativity, resulting in a comprehensive framework to describe the behavior of particles and fields in the universe. This synthesis laid the groundwork for the Standard Model of particle physics, which successfully describes the fundamental particles and forces of the universe, except for gravity. As quantum physics progressed, it also ventured into the realm of quantum computing and quantum information theory. The idea of using quantum bits, or qubits, as the fundamental unit of information introduced the possibility of exponentially faster computing and encryption-breaking algorithms. In recent years, quantum physicists have explored the exciting field of quantum optics, where they manipulate individual photons and explore the fascinating phenomena of quantum teleportation and quantum cryptography. These advancements have the potential to revolutionize communication and computation in the future. The birth *of quantum physics not only reshaped our understanding of the physical world but also challenged our philosophical and metaphysical assumptions. It revealed a reality that is inherently probabilistic, where particles can exist in multiple states simultaneously and where the act of measurement itself influences the outcome. This departure from classical determinism sparked debates about the nature of free will, consciousness, and the ultimate nature of reality. In summary, the pioneers and milestones of quantum physics have left an indelible mark on the landscape of science and our perception of the universe. From Planck's quantization to Schrödinger's wave equation, from Heisenberg's uncertainty principle to Einstein's photon hypothesis, the development of quantum theory reshaped the foundations of physics and opened the door to a new era of scientific exploration. Quantum physics has not only provided us with unprecedented technological advancements but has also challenged our worldview and invited us to ponder the profound mysteries of the quantum realm. It continues to be a frontier of discovery, offering glimpses into the deepest mysteries of the cosmos.*

Chapter 2: Waves and Particles: The Duality of Nature

Wave-particle duality is a fundamental concept in quantum physics that challenges our classical intuitions about the nature of particles and waves. This intriguing phenomenon suggests that particles, such as electrons and photons, exhibit both wave-like and particle-like properties depending on the context of the experiment. The concept of wave-particle duality emerged as a result of early experiments in the early 20th century, primarily involving the behavior of electrons. One of the key experiments that shed light on wave-particle duality was the double-slit experiment. In this experiment, a beam of particles, such as electrons or photons, is directed at a barrier with two closely spaced slits. When the particles pass through the slits and strike a screen on the other side, they create an interference pattern characteristic of waves. This interference pattern suggests that the particles exhibit wave-like behavior, with peaks and troughs where the waves reinforce or cancel each other out. However, the intriguing twist occurs when the particles are sent through the slits one at a time. Even when individual particles are sent through, they still create the same interference pattern over time, as if each particle is somehow interfering with itself. This phenomenon challenges the classical notion of particles as discrete, localized entities and suggests that they possess wave-like characteristics. Furthermore, the interference pattern fades away if the experimenters attempt to determine which slit each particle passes through. This suggests that the act of measurement or observation collapses the wave-like behavior, and the particles behave more like distinct, localized entities. The double-slit experiment is a clear illustration of the wave-particle duality concept. It demonstrates that particles can exhibit both wave-like interference and particle-like behavior, depending on whether they are observed or not. The wave-like behavior is associated with a particle's probability distribution, which describes the likelihood of finding the particle in a particular position. This distribution is represented by a wavefunction, a mathematical function that encodes the probability amplitudes associated with various positions. The square of the wavefunction's amplitude at a given point represents the probability of finding the particle at that position. The wavefunction evolves over time according to the Schrödinger equation, which describes how quantum states change with time. Wave-particle duality extends beyond the realm of electrons and photons and applies to other particles, such as protons, neutrons, and even larger molecules. However, the wavelength associated with a particle's wave-like behavior is inversely proportional to its momentum, meaning that larger particles have extremely tiny wavelengths and are, in practice, unlikely to exhibit noticeable wave behavior. While wave-particle duality challenges classical intuitions, it is essential for understanding the behavior of particles on the quantum scale. The probabilistic nature of quantum mechanics means that we can't predict the exact trajectory of a particle but *only its likelihood of being in a particular state or position. Wavefunctions and their associated probabilities provide a powerful framework for making predictions about the behavior of particles in quantum systems. The concept of wave-particle duality also has implications for the understanding of the behavior of matter and energy in the universe. In quantum field theory, which unifies quantum mechanics with special relativity, fields such as the electromagnetic field are quantized, meaning that they consist of discrete particles called quanta, or photons in the case of the electromagnetic field. These quanta exhibit both wave-like and particle-like behavior, just like electrons and photons. In particle physics, wave-particle duality is fundamental to our understanding of the behavior of subatomic particles and the fundamental forces that govern the universe. For example, the exchange of virtual particles, such as gluons in the strong force or W and Z bosons in the weak force, is mediated by particles that can be thought of as both waves and particles. Wave-particle duality also plays a crucial role in understanding the behavior of particles in accelerators like the Large Hadron Collider, where particles are accelerated to nearly the speed of light. In summary, wave-particle duality is a fundamental concept in quantum physics that challenges classical notions of particles and waves. It demonstrates that particles can exhibit both wave-like interference and particle-like behavior, depending on whether they are observed or not. This duality extends to particles of all sizes, from electrons to photons to subatomic particles, and is a central feature of quantum mechanics. Understanding wave-particle duality is essential for comprehending the behavior of matter and energy in the quantum world and has far-reaching implications in fields ranging from quantum field theory to particle physics. Experiments have played a crucial role in shaping our understanding of quantum physics, illuminating the strange and counterintuitive behavior of the quantum world. These experiments have challenged classical intuitions, expanded our knowledge, and provided valuable insights into the nature of reality at the smallest scales. One of the earliest experiments that laid the foundation for quantum mechanics was the blackbody radiation experiment. Max Planck, in 1900, proposed that energy is quantized into discrete packets, or quanta, to explain the spectral distribution of blackbody radiation. This concept marked a profound departure from classical physics and introduced the notion of quantization, which later became a cornerstone of quantum theory. Albert Einstein's work on the photoelectric effect in 1905 provided further experimental evidence for the existence of quanta. Einstein proposed that light consists of discrete packets of energy called photons, and his explanation of the photoelectric effect earned him the Nobel Prize in Physics in 1921. Another pivotal experiment was the double-slit experiment, which revealed the wave-particle duality of particles. When electrons or photons are directed at a barrier with two closely spaced slits, they create an interference pattern on the other side, suggesting wave-like behavior. However, even when* individual particles are sent through one at a time, they still create an interference pattern, challenging classical notions of particles as discrete entities. The double-slit experiment underscores the probabilistic nature of quantum measurements and the role of observation in collapsing the wavefunction. Werner Heisenberg's uncertainty principle, introduced in 1925, further emphasized the probabilistic nature of quantum mechanics. Heisenberg's principle states that it is impossible to simultaneously measure certain pairs of properties, such as a particle's position and momentum, with absolute precision. The act of measurement itself introduces uncertainty into the system, highlighting the inherent limitations of classical determinism. Niels Bohr's experiments with the hydrogen atom and his Bohr model in 1913 were instrumental in shaping the early understanding of quantum mechanics. Bohr's model incorporated the idea of quantized energy levels for electrons orbiting the nucleus, explaining the spectral lines of hydrogen. These experiments provided crucial insights into atomic structure and laid the groundwork for the development of quantum theory. In addition to these foundational experiments, quantum physics has been advanced by numerous experiments exploring various aspects of the quantum realm. Experiments with entangled particles have demonstrated the phenomenon of quantum entanglement, where two or more particles become intrinsically connected regardless of their spatial separation. Albert Einstein famously referred to this as "spooky action at a distance," highlighting the non-local and counterintuitive nature of entanglement. Experiments in quantum optics have allowed scientists to manipulate individual photons and explore phenomena such as quantum teleportation and quantum cryptography. These experiments have practical implications for the fields of communication and information security. Quantum computing experiments are pushing the boundaries of computation by harnessing the unique properties of qubits, the fundamental units of quantum information. Researchers are exploring the potential for exponentially faster computation and breakthroughs in solving complex problems. Experiments with ultra-cold atoms in traps have provided insights into the behavior of matter at extremely low temperatures, revealing phenomena such as Bose-Einstein condensates. Particle accelerators, like the Large Hadron Collider (LHC), have played a crucial role in experimental particle physics. The LHC, for example, was instrumental in the discovery of the Higgs boson, a fundamental particle responsible for imparting mass to other particles. Quantum experiments continue to explore the boundaries of our understanding, posing new questions and challenging existing theories. They provide a window into the mysterious and often counterintuitive behavior of the quantum world. Quantum mechanics is not just a theoretical framework but a set of principles that have been rigorously tested and confirmed through experiments. These experiments have not only expanded our scientific knowledge but have also led to technological advancements with practical applications. Quantum experiments have *enabled the development of technologies such as lasers, transistors, and nuclear reactors. They have paved the way for quantum information processing, with the potential to revolutionize computing and encryption. Quantum experiments have practical implications in fields ranging from telecommunications to medicine to materials science. Furthermore, they have sparked philosophical and metaphysical debates about the nature of reality, determinism, and the role of observation in shaping quantum outcomes. In summary, experiments have played a pivotal role in shaping our understanding of quantum physics and illuminating the peculiar behavior of the quantum world. From blackbody radiation to the double-slit experiment, from the photoelectric effect to quantum entanglement, these experiments have challenged classical intuitions and expanded our knowledge. Quantum experiments have practical applications in technology and science while also sparking profound philosophical questions about the nature of reality.*

Chapter 3: Quantum Superposition: The Art of Being in Two Places at Once

Exploring superposition is a fundamental aspect of understanding quantum physics and its unique properties. Superposition is a concept that allows quantum particles to exist in multiple states simultaneously, in contrast to classical physics, where objects have definite properties. This intriguing phenomenon is a cornerstone of quantum mechanics and plays a crucial role in many quantum applications. Superposition arises from the mathematical description of quantum states using wavefunctions. A quantum state, represented by a wavefunction, encodes information about a particle's properties, such as position, momentum, or spin. In a classical system, a particle's properties are well-defined, and we can specify its state precisely. However, in quantum mechanics, a particle's state can be in a superposition of multiple possible states. For example, consider an electron's spin, which can be either "up" or "down" in a magnetic field. In classical physics, the electron's spin would be either "up" or "down" at any given moment. In contrast, in a quantum system, the electron's spin can exist in a superposition of "up" and "down" states until measured, whereupon it collapses into one of those states. This means that before measurement, the electron's spin is not definitively "up" or "down" but has a certain probability of being either. Superposition extends beyond spin states and applies to various quantum observables, such as the position and momentum of particles. In the double-slit experiment, particles like electrons or photons exhibit superposition when they pass through two slits simultaneously, creating an interference pattern on the screen. This interference pattern results from the superposition of possible paths the particles can take. In essence, each particle explores multiple paths at once, interfering with itself and creating the observed pattern. Superposition also plays a central role in quantum computing, where quantum bits, or qubits, can exist in multiple states simultaneously. This property allows quantum computers to perform certain calculations exponentially faster than classical computers. For instance, Shor's algorithm, a quantum algorithm, can factor large numbers efficiently, posing a significant threat to classical encryption methods. Superposition is harnessed in quantum algorithms like Grover's algorithm to search unsorted databases more efficiently. Quantum superposition also has practical applications in quantum cryptography and secure communication. Quantum key distribution relies on the principle of superposition to secure communication channels against eavesdropping. Entanglement, another fundamental quantum phenomenon, is closely related to superposition. When two particles become entangled, their properties are linked, and they exist in a joint superposition of states. Measuring one particle instantly determines the state of the other, regardless of the distance separating them. This "spooky action at a distance," as Einstein called it, *highlights the non-local nature of quantum superposition and entanglement. Superposition is not limited to particles but extends to quantum systems of all sizes, including atoms and molecules. For example, in nuclear magnetic resonance (NMR) spectroscopy, superposition is used to study the properties of molecules. Superposition of nuclear spins in a magnetic field allows researchers to gather information about molecular structures and dynamics. In the field of quantum optics, superposition is exploited to manipulate individual photons and study their properties. Quantum teleportation experiments involve entangling two photons and placing one in a superposition of states, enabling the transfer of quantum information over long distances. Superposition is not without its challenges and complexities. Maintaining a quantum system in a superposition state typically requires careful control and isolation from external influences, which can lead to decoherence and the collapse of the superposition. Noise and environmental factors can cause superposition to decay quickly, limiting its usefulness in practical applications. Scientists and engineers are actively researching ways to mitigate decoherence and extend the duration of superposition in quantum systems. In summary, exploring superposition in quantum systems is essential for understanding the unique properties and potential applications of quantum mechanics. Superposition allows quantum particles and systems to exist in multiple states simultaneously until measured, opening the door to groundbreaking technologies like quantum* computing and secure communication. While superposition presents challenges related to decoherence, ongoing research aims to harness its power for practical use in various fields, from cryptography to molecular spectroscopy. This concept continues to captivate the imagination of scientists and researchers, as it offers unprecedented opportunities for exploring the mysteries of the quantum realm and pushing the boundaries of what is possible in the world of technology and science. Quantum superposition, a fundamental concept in quantum mechanics, challenges our classical intuitions and has far-reaching implications in the real world. While superposition may seem like an abstract concept, it plays a crucial role in various practical applications and technologies. One of the most well-known applications of superposition is in quantum computing, where quantum bits, or qubits, can exist in multiple states simultaneously. This property allows quantum computers to perform certain calculations exponentially faster than classical computers. Shor's algorithm, for example, exploits superposition to efficiently factor large numbers, posing a significant threat to classical encryption methods. Quantum algorithms like Grover's algorithm leverage superposition to search unsorted databases more efficiently, offering advantages in data analysis and optimization problems. Superposition is not limited to the realm of computing; it also plays a pivotal role in quantum cryptography. Quantum key distribution, based on the principles of superposition and entanglement, enables secure communication by detecting any eavesdropping attempts. The use of superposition in quantum encryption methods ensures that intercepted data cannot be deciphered without altering the quantum states, alerting users to potential security breaches. Beyond computing and cryptography, superposition has practical applications in quantum sensors and metrology. Quantum sensors, such as atomic clocks and magnetometers, utilize the sensitivity of quantum states in superposition to measure physical quantities with unprecedented precision. For instance, atomic clocks rely on the superposition of quantum states in atoms to achieve remarkable accuracy in timekeeping, which is crucial for global positioning systems (GPS) and synchronization in telecommunications. Quantum magnetometers leverage superposition to detect minute magnetic fields, making them valuable tools in various fields, including geophysics and medical imaging. Superposition also plays a role in quantum-enhanced imaging techniques, enabling the detection of faint signals and improving the resolution of imaging devices. In the field of quantum optics, superposition is exploited to manipulate individual photons and study their properties. Quantum teleportation experiments involve entangling two photons and placing one in a superposition of states, enabling the transfer of quantum information over long distances. Quantum cryptography and secure communication systems leverage the principles of superposition to protect sensitive data from potential eavesdroppers. These real-world applications highlight the practical significance of superposition in modern *technology and science. Superposition is not without its challenges, particularly concerning decoherence. Decoherence occurs when a quantum system interacts with its environment, causing the superposition to collapse into a classical state. To harness the power of superposition effectively, researchers must address the issue of decoherence and find ways to extend the duration of quantum states. Various techniques, such as error-correcting codes and quantum error correction, are being developed to mitigate the effects of decoherence and enhance the stability of quantum systems. Superposition has also inspired research in quantum simulations, where quantum computers simulate complex quantum systems that are challenging to study using classical methods. Quantum simulators use the principles of superposition to explore phenomena in condensed matter physics, quantum chemistry, and materials science. These simulations offer insights into the behavior of quantum systems and can lead to the discovery of new materials with unique properties. In the realm of quantum communication, superposition enables the development of secure communication protocols that are immune to hacking attempts. Quantum key distribution systems rely on the superposition of quantum states to generate and exchange cryptographic keys, ensuring the confidentiality of transmitted information. These systems hold great promise for secure communication in an increasingly interconnected world. Superposition also has implications in the study of quantum mechanics itself. Experiments involving superposition provide valuable insights into the fundamental nature of quantum particles and their behavior. The double-slit experiment, a classic example, reveals the wave-particle duality of particles and demonstrates superposition in action. Particles like electrons or photons exhibit superposition when they pass through two slits simultaneously, creating an interference pattern on the screen. Even when sent through one at a time, they still create an interference pattern, challenging classical notions of particles as discrete entities. This experiment underscores the probabilistic nature of quantum measurements and the role of observation in collapsing the wavefunction. The phenomenon of quantum entanglement, closely related to superposition, raises profound questions about the nature of reality. Entangled particles exist in a joint superposition of states, and measuring one particle instantly determines the state of the other, regardless of the distance separating them. This "spooky action at a distance," as Einstein called it, highlights the non-local and counterintuitive nature of quantum superposition and entanglement. Superposition continues to captivate the imagination of scientists and researchers, offering unprecedented opportunities for exploring the mysteries of the quantum realm. It challenges our classical intuitions, expands our technological capabilities, and pushes the boundaries of what is possible in the world of technology and science. As researchers continue to advance our understanding of superposition and address the challenges it presents, we can expect even more groundbreaking applications and discoveries in the future. In summary, quantum superposition is not just a theoretical concept; it has a profound impact on the real world, driving advancements in quantum computing, cryptography, sensors, imaging, and more. While decoherence remains a challenge, ongoing research and innovation are opening up new possibilities for harnessing the power of superposition in practical applications and scientific exploration. Superposition is at the heart of the quantum revolution, reshaping our understanding of the quantum realm and shaping the future of technology and science.*

Chapter 4: The Uncertainty Principle: Limits of Precision

Heisenberg's Uncertainty Principle, a fundamental concept in quantum mechanics, asserts a fundamental limit on our ability to simultaneously know certain pairs of properties of a particle with absolute precision. Named after the German physicist Werner Heisenberg, this principle challenges classical intuitions about the determinacy of the physical world. The Uncertainty Principle, formulated in 1927, states that there is an inherent trade-off between the precision with which we can know a particle's position and its momentum. In other words, the more accurately we measure a particle's position, the less precisely we can determine its momentum, and vice versa. This principle arises from the wave-particle duality inherent in quantum mechanics. In classical physics, it is assumed that both position and momentum can be determined with arbitrary precision, and the properties of particles are deterministic. However, Heisenberg's Uncertainty Principle introduces an element of indeterminacy into the quantum world. To illustrate this principle, consider an experiment involving the measurement of an electron's position and momentum. If we use a high-energy photon to measure the electron's position very precisely, the impact of the photon on the electron's momentum becomes significant, causing a substantial change in the electron's momentum. Conversely, if we use a low-energy photon to minimize the disturbance to the electron's momentum, we obtain a less precise measurement of its position. This trade-off is a fundamental limitation of quantum mechanics and implies that there are inherent limits to the precision of measurements in the quantum world. The Uncertainty Principle is not just a limitation of measurement technology; it reflects the intrinsic nature of quantum particles. The uncertainty in a particle's position and momentum is related to the spread or "width" of its wavefunction, a mathematical description of the probability distribution of finding the particle in different positions. The narrower the wavefunction (indicating a more precise position measurement), the wider it becomes in momentum space (indicating a less precise momentum measurement), and vice versa. Heisenberg's Uncertainty Principle has profound implications for our understanding of quantum mechanics and the behavior of particles. It challenges the classical notion of deterministic trajectories and emphasizes the probabilistic nature of quantum systems. In a sense, it suggests that particles do not possess well-defined properties until they are measured, and even then, there is a fundamental limit to the precision of our knowledge. The Uncertainty Principle also has practical implications in various fields. One of the most notable applications is in nuclear physics and the study of subatomic particles. In particle accelerators like the Large Hadron Collider (LHC), physicists accelerate particles to extremely high energies and collide them to study the fundamental building blocks of matter. The Uncertainty Principle *plays a critical role in these experiments, as the precise measurement of a particle's position and momentum is essential for understanding its behavior. Additionally, the Uncertainty Principle is central to the development of technologies such as electron microscopes and scanning tunneling microscopes, which rely on the precise measurement of particles at the nanoscale. In quantum mechanics, the concept of wavefunctions is used to describe the probabilistic nature of particles. Wavefunctions encode information about a particle's position, momentum, and other properties. The Uncertainty Principle is mathematically expressed using the standard deviations of the position and momentum operators in terms of the wavefunction. It is represented as $\Delta x \Delta p \geq \hbar/2$, where Δx is the uncertainty in position, Δp is the uncertainty in momentum, and $\hbar$ (h-bar) is the reduced Planck constant, a fundamental constant of nature. The Uncertainty Principle has been experimentally confirmed in numerous experiments, providing empirical evidence for its validity. For example, experiments involving electron diffraction and the behavior of particles in magnetic fields have consistently demonstrated the trade-off between position and momentum measurements, consistent with the Uncertainty Principle. The Uncertainty Principle also extends beyond the position-momentum pair and applies to other pairs of complementary observables in quantum mechanics. For instance, it applies to pairs such as energy and time or angular position and angular momentum. In the case of energy and time, the more precisely we know the energy of a quantum system, the* *less precisely we can determine the time at which a particular event occurred, and vice versa. Heisenberg's Uncertainty Principle has implications for our understanding of quantum systems and the limitations of classical intuition in the quantum realm. It challenges the idea of a deterministic universe and underscores the probabilistic nature of quantum mechanics. The principle has practical applications in physics, technology, and quantum experiments, and its mathematical formulation provides a fundamental constraint on the precision of measurements in the quantum world. While the Uncertainty Principle introduces uncertainty into the behavior of quantum particles, it also opens the door to the exploration of the intriguing and counterintuitive aspects of the quantum world. It invites us to embrace the inherent uncertainty of the quantum realm and to appreciate the profound insights it offers into the nature of reality. In summary, Heisenberg's Uncertainty Principle is a fundamental concept in quantum mechanics that places limits on our ability to simultaneously know certain pairs of properties of a particle with absolute precision. It challenges classical determinism, emphasizing the probabilistic nature of quantum systems and their inherent uncertainty. The principle has practical applications in various fields, and its mathematical formulation provides a fundamental constraint on the precision of measurements in the quantum world. While the Uncertainty Principle introduces uncertainty, it also opens the door to a deeper understanding of the quantum realm and the mysteries it holds.*

The consequences and implications of Heisenberg's Uncertainty Principle reach far beyond the realm of quantum mechanics, extending into various aspects of physics, philosophy, and technology. This fundamental principle, which limits our ability to simultaneously know certain pairs of properties of a particle with absolute precision, has profound consequences for our understanding of the physical world. One of the immediate consequences of the Uncertainty Principle is the introduction of indeterminacy into the behavior of quantum particles. In classical physics, it was often assumed that particles have well-defined properties, such as definite positions and momenta. However, the Uncertainty Principle suggests that until a measurement is made, a quantum particle's properties exist in a superposition of possible values. This probabilistic nature challenges our classical intuitions and underscores the inherently uncertain character of the quantum realm. The Uncertainty Principle also implies that particles do not have determinate trajectories, as classical physics would suggest. Instead, a particle's motion is described by a probability distribution, represented by its wavefunction, which encodes the likelihood of finding the particle in different positions and with different momenta. This probabilistic description is a departure from classical determinism and has significant implications for our understanding of causality in the quantum world. Furthermore, the Uncertainty Principle has philosophical implications, sparking debates about the nature of reality and the role of measurement in shaping quantum outcomes.

Some interpretations of quantum mechanics, such as the Copenhagen interpretation, emphasize the role of the observer in collapsing the wavefunction and determining the outcome of a measurement. This raises questions about the nature of reality when it is not being observed and the possibility of objective reality independent of observation. The philosophical implications of the Uncertainty Principle continue to be a subject of ongoing discussion among physicists and philosophers. In addition to its philosophical implications, the Uncertainty Principle has practical consequences for experiments and measurements in the quantum world. It imposes fundamental limits on the precision with which certain pairs of properties can be simultaneously known. For example, in particle physics experiments, the Uncertainty Principle places constraints on the precision of measurements of a particle's position and momentum. To improve the precision of one measurement, such as position, the uncertainty in the complementary property, momentum, must increase. This limitation is not due to technological constraints but is an intrinsic property of quantum systems. As a result, researchers must carefully consider the trade-offs between the precision of measurements and the associated uncertainties when designing experiments. The Uncertainty Principle also plays a crucial role in the development of quantum technologies and quantum computing. Quantum computers, which leverage the principles of superposition and entanglement, can perform certain calculations much faster than classical computers. However, the

Uncertainty Principle sets a limit on the precision of measurements and operations in quantum computing systems. Efforts to mitigate the effects of decoherence, which causes the collapse of superposition, are essential to building practical and reliable quantum computers. Furthermore, the Uncertainty Principle has practical applications in fields such as quantum cryptography and secure communication. Quantum key distribution systems use the principles of superposition and entanglement to generate cryptographic keys that are immune to eavesdropping attempts. The Uncertainty Principle ensures that any attempt to intercept quantum-encoded information will disturb the quantum states, alerting users to potential security breaches. Quantum cryptography holds promise for enhancing the security of communication in an increasingly interconnected world. The Uncertainty Principle also has implications for the study of subatomic particles and the behavior of matter and energy in the universe. In particle physics experiments, the Uncertainty Principle places constraints on the precision with which we can know the properties of particles at the quantum level. This limitation is fundamental and affects our understanding of the behavior of particles in high-energy collisions. The behavior of particles in accelerators like the Large Hadron Collider (LHC) is governed by quantum mechanics, and the Uncertainty Principle plays a critical role in the interpretation of experimental results. Furthermore, the Uncertainty Principle has implications for our understanding of quantum field theory, which describes the behavior of fields such as the electromagnetic field, and the exchange of virtual particles. In the study of cosmology and the early universe, the Uncertainty Principle plays a role in understanding the behavior of matter and energy in extreme conditions. As the universe expands and cools, the properties of particles and fields can be influenced by the principles of quantum mechanics, including the Uncertainty Principle. In summary, Heisenberg's Uncertainty Principle has consequences and implications that reach far beyond the realm of quantum mechanics. It introduces indeterminacy into the behavior of quantum particles, challenges classical intuitions about determinism, and has philosophical implications for the nature of reality and the role of measurement. The Uncertainty Principle also imposes fundamental limits on the precision of measurements in quantum systems, influencing experimental design and technology development. Practical applications in quantum computing and cryptography highlight its relevance in modern technology. In the study of subatomic particles and the behavior of matter and energy in the universe, the Uncertainty Principle plays a critical role in our understanding of the quantum world. It continues to shape our view of the physical world, inspiring research and exploration into the mysteries of the quantum realm.

Chapter 5: Entanglement: Spooky Action at a Distance

The EPR paradox, named after its authors Albert Einstein, Boris Podolsky, and Nathan Rosen, is a foundational concept in the field of quantum mechanics. This paradox emerged in 1935 as a thought experiment designed to challenge the completeness and implications of quantum theory. Einstein, Podolsky, and Rosen sought to demonstrate what they perceived as the shortcomings of quantum mechanics, particularly its prediction of entanglement, a phenomenon they found deeply unsettling. Entanglement is a central feature of quantum physics, and it refers to the phenomenon where two or more particles become intrinsically connected, regardless of the distance that separates them. In an entangled state, the properties of one particle are correlated with the properties of another, such that measuring one particle instantaneously determines the state of the other, even if they are light-years apart. Einstein famously referred to this phenomenon as "spooky action at a distance" and was troubled by the idea that distant particles could instantaneously influence each other. The EPR paradox was formulated to challenge this notion of entanglement and its implications. The thought experiment proposed by EPR began with the concept of a pair of particles, such as electrons, that are prepared in an entangled state. According to quantum mechanics, the properties of these particles, such as their positions and momenta, are described by wavefunctions that are entangled. If one particle is measured, and its properties are determined, the other particle's properties are immediately known, even if it is far away. Einstein, Podolsky, and Rosen argued that this meant that the properties of the distant particle must have been predetermined even before the measurement took place. They claimed that the theory was incomplete because it did not provide a complete description of the properties of the particles without measurement. This, they believed, violated the principles of classical physics and determinism. Einstein, Podolsky, and Rosen asserted that the particles must have hidden variables, which would determine their properties independently of any measurement. They argued that quantum mechanics was incomplete because it did not provide a full description of these hidden variables. Einstein, in particular, held a strong belief in the existence of a deterministic and complete theory that would explain all physical phenomena. He hoped that the EPR paradox would demonstrate the inadequacy of quantum mechanics and inspire the development of such a theory. However, in 1964, physicist John Bell formulated a set of mathematical inequalities, known as Bell's inequalities, which could be tested experimentally. Bell's inequalities provided a way to distinguish between the predictions of classical physics, which assumed the existence of hidden variables, and the predictions of quantum mechanics. Experiments based on Bell's inequalities have consistently shown that the predictions of quantum mechanics are correct, while those of classical physics are not. These experiments have provided strong evidence in favor of entanglement and have refuted the idea of hidden variables determining the properties of particles. In other words, entanglement is not the result of pre-determined properties but is a genuine feature of the quantum world. The EPR paradox, rather than disproving quantum mechanics, highlighted the profound and counterintuitive nature of quantum entanglement. It demonstrated that quantum theory's predictions about entanglement were fundamentally different from those of classical physics, challenging our classical intuitions about the determinacy and separability of physical systems. The implications of entanglement are far-reaching and have practical applications in fields such as quantum cryptography and quantum computing. Quantum cryptography leverages entanglement to create secure communication protocols that are resistant to eavesdropping. Because any attempt to intercept the quantum-encoded information would disturb the entangled states, any eavesdropping attempt would be detected. Quantum computing, on the other hand, exploits entanglement to perform certain calculations exponentially faster than classical computers. Quantum bits, or qubits, in a superposition of states can perform complex calculations by simultaneously exploring multiple possibilities. The phenomenon of entanglement has also led to groundbreaking experiments and insights into the foundations of quantum mechanics. Experiments testing Bell's inequalities continue to provide support for the validity of quantum theory and *the reality of entanglement. These experiments have demonstrated that entanglement violates the principles of classical physics and have further reinforced the importance of quantum entanglement in the quantum world. Additionally, entanglement has led to the development of quantum teleportation, a process that allows the transfer of quantum information from one location to another, albeit without physical transport of particles. Quantum teleportation relies on entanglement to achieve secure and instantaneous information transfer, with potential applications in quantum communication and information processing. In summary, the EPR paradox and the subsequent experiments based on Bell's inequalities have played a pivotal role in our understanding of entanglement and the counterintuitive nature of quantum mechanics. Entanglement is a fundamental feature of the quantum world, challenging classical intuitions about the determinacy and separability of physical systems. It has practical applications in quantum cryptography and quantum computing, offering new possibilities for secure communication and computational speedup. Furthermore, entanglement has inspired groundbreaking experiments that continue to shape our understanding of the quantum realm and the profound implications of quantum mechanics. The EPR paradox, initially formulated as a challenge to quantum mechanics, ultimately revealed the fundamental and unique nature of entanglement, reinforcing the central role of quantum phenomena in the physical world. Quantum entanglement is one of the most intriguing and perplexing phenomena in the field of quantum physics. It was famously described by Albert Einstein as "spooky action at a distance" due to its seemingly paradoxical nature. Entanglement occurs when two or more particles become correlated or linked in such a way that the state of one particle is dependent on the state of another, regardless of the distance that separates them. This means that measuring one particle instantaneously determines the state of the other, even if they are light-years apart. The concept of entanglement challenges our classical intuitions about the separability of physical systems and has profound implications for our understanding of quantum mechanics. One of the key features of entanglement is that it is a non-local phenomenon. In classical physics, information cannot travel faster than the speed of light, and the properties of distant objects are not instantaneously connected. However, entangled particles violate this classical boundary and demonstrate a form of non-locality. This means that changes in the state of one particle can affect the state of the other particle, regardless of the spatial separation between them. The phenomenon of quantum entanglement was first proposed as a thought experiment by Albert Einstein, Boris Podolsky, and Nathan Rosen in their 1935 paper, often referred to as the EPR paradox. Einstein, Podolsky, and Rosen designed this thought experiment to challenge the completeness and implications of quantum theory. They argued that if two particles were entangled, as predicted by quantum mechanics, then measuring one particle's properties would instantaneously determine the properties of the other particle, even if it were located far away. This, they believed, violated the principles of classical determinism and causality. Einstein, in particular, was uncomfortable with the idea that distant particles could have instantaneous influence over each other, and he proposed that there must be "hidden variables" that determined the properties of particles independently of measurement. The EPR paradox aimed to demonstrate that quantum mechanics was an incomplete theory and that a more fundamental theory with hidden variables was needed. However, in 1964, physicist John Bell formulated a set of mathematical inequalities, known as Bell's inequalities, which provided a way to test the predictions of quantum mechanics against those of classical physics with hidden variables. Experiments based on Bell's inequalities have consistently shown that the predictions of quantum mechanics are correct, while those of classical physics with hidden variables are not. This experimental evidence supports the existence of entanglement and non-locality in the quantum world. One of the most famous experiments testing Bell's inequalities is the Aspect experiment, conducted by Alain Aspect in the 1980s. Aspect's experiments involved measuring the polarization of entangled photons emitted from a single source. The results of these experiments violated Bell's inequalities and provided strong evidence for the existence of quantum entanglement. Entanglement is not limited to particles of a specific type; it can occur with various quantum systems, including electrons, photons, and atoms. For example, consider the case of entangled electrons. Two electrons can be placed in an entangled state in which their spins are correlated. If one electron is measured to have an "up" spin, the other electron will have a "down" spin, and vice versa, even if they are separated by large distances. In the case of entangled photons, their polarizations can be correlated, and measuring one photon's polarization will determine the polarization of the other, regardless of the separation between them. Entanglement also plays a significant role in quantum teleportation, a process that allows the transfer of quantum information from one location to another without physical transport of particles. Quantum teleportation relies on entanglement to achieve secure and instantaneous information transfer, with potential applications in quantum communication and information processing. The phenomenon of entanglement raises several intriguing questions about the nature of reality and the role of measurement in quantum mechanics. Some interpretations of quantum mechanics, such as the Copenhagen interpretation, emphasize the role of the observer in collapsing the wavefunction and determining the outcome of a measurement. This has led to debates about the nature of reality when it is not being observed and the possibility of an objective reality independent of observation. The philosophical implications of entanglement continue to be a subject of ongoing discussion among physicists and philosophers. Furthermore, entanglement has practical applications in various fields. Quantum cryptography leverages entanglement to create secure communication protocols that are resistant to eavesdropping. Any attempt to intercept the quantum-encoded information would disturb the entangled states, alerting users to potential security breaches. Quantum computing, on the other hand, exploits entanglement to perform certain calculations exponentially faster than classical computers. Quantum bits, or qubits, in a superposition of states can perform complex calculations by simultaneously exploring multiple possibilities. In summary, quantum entanglement is a remarkable and counterintuitive phenomenon in the quantum world. It challenges classical intuitions about the separability of physical systems, introduces non-locality into quantum mechanics, and has profound philosophical implications. Experiments based on Bell's inequalities have provided strong empirical evidence for the existence of entanglement, reinforcing its fundamental role in the quantum realm. Entanglement has practical applications in quantum cryptography, quantum computing, and quantum teleportation, offering new possibilities for secure communication and computational speedup. The concept of entanglement continues to captivate the imagination of scientists, inspiring research into the mysteries of the quantum world and pushing the boundaries of our understanding of the physical universe.*

Chapter 6: Quantum Mechanics in Everyday Life

Quantum mechanics, the branch of physics that describes the behavior of matter and energy at the smallest scales, has had a profound impact on technology and has given rise to a new era of innovation. One of the most transformative applications of quantum mechanics is in the field of quantum computing. Quantum computers, based on the principles of superposition and entanglement, have the potential to perform certain calculations exponentially faster than classical computers. This extraordinary computational power has far-reaching implications for fields such as cryptography, materials science, and optimization. In quantum computing, the fundamental unit of information is the quantum bit, or qubit, which can exist in a superposition of states. This means that a qubit can represent both 0 and 1 simultaneously, allowing quantum computers to explore multiple possibilities in parallel. Shor's algorithm, for example, exploits this property to efficiently factor large numbers, posing a significant threat to classical encryption methods. Quantum algorithms like Grover's algorithm leverage superposition to search unsorted databases more efficiently, offering advantages in data analysis and optimization problems. Quantum computing is still in its infancy, with many technical challenges to overcome, but it holds immense promise for solving complex problems that are currently beyond the reach *of classical computers. Another groundbreaking application of quantum mechanics is in quantum cryptography. Quantum key distribution (QKD) systems use the principles of superposition and entanglement to generate cryptographic keys that are immune to eavesdropping attempts. The Uncertainty Principle ensures that any attempt to intercept quantum-encoded information would disturb the quantum states, alerting users to potential security breaches. Quantum cryptography offers a new level of security for communication in an increasingly interconnected world and has the potential to revolutionize the field of cybersecurity. Quantum mechanics has also paved the way for the development of quantum sensors and metrology. These sensors utilize the sensitivity of quantum states to measure physical quantities with unprecedented precision. For instance, atomic clocks rely on the superposition of quantum states in atoms to achieve remarkable accuracy in timekeeping, which is crucial for global positioning systems (GPS) and synchronization in telecommunications. Quantum magnetometers leverage superposition to detect minute magnetic fields, making them valuable tools in various fields, including geophysics and medical imaging. Superposition also plays a role in quantum-enhanced imaging techniques, enabling the detection of faint signals and improving the resolution of imaging devices. In the field of quantum optics, superposition is exploited to manipulate individual photons and study their properties. Quantum teleportation experiments involve entangling two photons and placing one in a superposition of states, enabling the transfer of quantum information over long distances. Quantum cryptography and secure communication systems leverage the principles of superposition to protect sensitive data from potential eavesdroppers. These real-world applications highlight the practical significance of superposition in modern technology and science. Quantum mechanics has also found applications in materials science and nanotechnology. The behavior of quantum systems at the nanoscale can be harnessed to create novel materials with unique properties. For example, quantum dots are nanoscale semiconductor particles that exhibit quantum mechanical behavior. They have been used in the development of high-efficiency solar cells, quantum dot displays, and quantum dot-based sensors. Quantum mechanical simulations are employed to understand and predict the properties of these nanomaterials, facilitating their design and optimization. Moreover, the principles of quantum mechanics underpin the operation of scanning tunneling microscopes (STMs) and atomic force microscopes (AFMs), which are essential tools for imaging and manipulating individual atoms and molecules. STMs, in particular, rely on the quantum mechanical phenomenon of tunneling to create atomic-scale images of surfaces and manipulate individual atoms. The principles of quantum mechanics have also revolutionized the field of quantum chemistry. Quantum chemical calculations, based on the Schrödinger equation and the principles of superposition, are used to study the electronic structure and properties of* molecules and materials. These calculations provide valuable insights into chemical reactions, material properties, and molecular interactions, accelerating the discovery and development of new drugs, catalysts, and materials. In addition to its applications in technology and science, quantum mechanics has had a profound impact on our understanding of the fundamental laws of the universe. Quantum field theory, which combines quantum mechanics with special relativity, provides a framework for describing the behavior of particles and fields in the universe. It has been highly successful in explaining the behavior of subatomic particles and their interactions through the exchange of virtual particles. Quantum field theory is the basis of the Standard Model of particle physics, which describes the electromagnetic, weak, and strong nuclear forces and the particles that mediate them. Furthermore, the principles of quantum mechanics have implications for our understanding of the fundamental nature of reality. Experiments involving superposition provide valuable insights into the probabilistic nature of quantum measurements and the role of observation in collapsing the wavefunction. The double-slit experiment, for example, reveals the wave-particle duality of particles and demonstrates superposition in action. Particles like electrons or photons exhibit superposition when they pass through two slits simultaneously, creating an interference pattern on the screen. Even when sent through one at a time, they still create an interference pattern, challenging classical notions of particles as discrete entities. This experiment underscores the probabilistic *nature of quantum measurements and the role of observation in collapsing the wavefunction. The phenomenon of quantum entanglement, closely related to superposition, raises profound questions about the nature of reality. Entangled particles exist in a joint superposition of states, and measuring one particle instantly determines the state of the other, regardless of the distance separating them. This "spooky action at a distance," as Einstein called it, highlights the non-local and counterintuitive nature of quantum superposition and entanglement. Superposition and entanglement continue to captivate the imagination of scientists and researchers, offering unprecedented opportunities for exploring the mysteries of the quantum realm. They challenge our classical intuitions, expand our technological capabilities, and push the boundaries of what is possible in the world of technology and science. As researchers continue to advance our understanding of superposition and entanglement and harness their power for practical applications, quantum mechanics will undoubtedly play an increasingly central role in shaping the future of technology and our understanding of the universe.*

The concepts and principles of quantum mechanics, though often associated with the microscopic realm of atoms and subatomic particles, have a surprising impact on our daily lives. Quantum mechanics underlies the behavior of matter and energy at the smallest scales, but its effects can extend into the macroscopic world we inhabit. One of the most ubiquitous applications of quantum mechanics is in the field of electronics. Transistors, the fundamental building blocks of modern electronic devices, rely on the quantum mechanical phenomenon of electron tunneling. In a transistor, electrons can move from one region to another by tunneling through a barrier, which is a quantum process governed by the probabilistic nature of quantum mechanics. This allows transistors to amplify and switch electronic signals, forming the basis of computers, smartphones, and countless other electronic devices that permeate our lives. Without quantum mechanics, these devices would not exist in their current form. Quantum mechanics also plays a crucial role in the design and operation of semiconductor lasers, which are used in various applications, including optical communication, barcode scanners, and laser printers. The principles of quantum mechanics govern the energy levels and transitions of electrons in semiconductors, allowing them to emit coherent and highly focused light. These lasers are essential in modern technology and are a testament to the practical impact of quantum mechanics on our daily lives. In the realm of medical imaging, quantum mechanics is at the heart of magnetic resonance imaging (MRI). MRI machines use the quantum property of nuclear spin to create detailed images of the internal structures of the human body. In an MRI scan, the alignment and relaxation of atomic nuclei in a magnetic field are manipulated and measured, providing valuable diagnostic information without the need for ionizing radiation. This application of quantum mechanics has revolutionized medical diagnosis and is an example of how quantum principles are directly beneficial to our health and well-being. Quantum mechanics also plays a role in the development of new materials with remarkable properties. Superconductors, for instance, are materials that exhibit zero electrical resistance at low temperatures. The behavior of superconductors is described by the principles of quantum mechanics, specifically the formation of Cooper pairs of electrons and their condensation into a quantum state. Superconductors have applications in the creation of powerful magnets for medical imaging and the generation of electricity with minimal loss, potentially revolutionizing energy transmission and storage. Furthermore, quantum dots, nanoscale semiconductor particles that exhibit quantum mechanical behavior, are being explored for their applications in displays, lighting, and solar cells. The unique properties of quantum dots, such as size-dependent electronic and optical properties, have the potential to enhance the efficiency of energy conversion and storage devices. In the realm of energy production, quantum mechanics contributes to the development of advanced photovoltaic materials that can capture and convert sunlight into electricity more efficiently. Quantum mechanical simulations are employed to understand the electronic and optical properties of materials, guiding the design of next-generation solar cells. Moreover, quantum mechanics plays a critical role in the development of advanced materials with specific electronic, optical, and mechanical properties. The principles of quantum mechanics are used to model the behavior of atoms and molecules, enabling the discovery and design of new materials with tailored properties. This has applications in industries ranging from aerospace to telecommunications, where materials with exceptional strength, conductivity, or transparency are required. The impact of quantum mechanics on our daily lives extends to the field of chemistry. Chemical reactions and molecular interactions are governed by quantum mechanics. Understanding the electronic structure and behavior of molecules at the quantum level is essential for drug discovery, materials design, and environmental analysis. Quantum chemical calculations, which solve the Schrödinger equation for molecular systems, provide insights into chemical reactions, molecular properties, and the behavior of complex molecules. These calculations are crucial for predicting the behavior of molecules in diverse applications, from the development of pharmaceuticals to the design of catalysts for industrial processes. The principles of quantum mechanics also underpin the field of quantum chemistry, which explores the behavior of molecules and chemical reactions using quantum mechanical models. This field has revolutionized our understanding of chemical processes and has practical applications in drug design, materials science, and environmental chemistry. In the world of telecommunications, the principles of quantum mechanics enable secure communication through quantum key distribution (QKD). QKD systems use the quantum properties of entanglement and superposition to generate cryptographic keys that are immune to *eavesdropping attempts. The Uncertainty Principle ensures that any attempt to intercept quantum-encoded information would disturb the quantum states, alerting users to potential security breaches. Quantum cryptography offers a new level of security for communication, addressing the growing concerns about data privacy and security in our interconnected world. Furthermore, quantum mechanics has practical applications in quantum sensors and metrology, where the sensitivity of quantum states is harnessed to measure physical quantities with unprecedented precision. Quantum sensors, such as atomic clocks and magnetometers, have applications in GPS systems, telecommunications, and geophysical exploration. Their high precision relies on the quantum properties of atoms and particles, making them indispensable tools in modern technology. Quantum mechanics also plays a role in the field of quantum optics, where the behavior of photons and their interactions with matter are explored. Quantum optics has applications in imaging, quantum communication, and the development of quantum technologies. One of the most fascinating aspects of quantum mechanics is its potential to enable technologies that harness the principles of superposition and entanglement for quantum computing. Quantum computers have the potential to solve complex problems in fields such as cryptography, materials science, and optimization that are currently beyond the capabilities of classical computers. Researchers and companies are actively working on building practical quantum computers that could revolutionize industries and scientific research. In summary, the principles of quantum mechanics have a profound and far-reaching impact on our daily lives, influencing a wide range of technologies and applications. From the transistors in our electronic devices to the secure communication provided by quantum cryptography, from the precision of medical imaging to the development of advanced materials, quantum mechanics is an essential part of the modern technological landscape. As our understanding of quantum mechanics deepens and our ability to harness its principles advances, we can expect even more transformative innovations that will continue to shape the way we live, work, and interact with the world around us.*

Chapter 7: The Schrödinger Equation: The Heart of Quantum Physics

The Schrödinger equation is a fundamental equation in quantum mechanics that describes the behavior of quantum systems. Named after Austrian physicist Erwin Schrödinger, who formulated it in 1926, the equation is central to our understanding of the quantum world. At its core, the Schrödinger equation is a wave equation that governs how the wavefunction of a quantum system evolves over time. The wavefunction, denoted by the symbol Ψ (psi), is a mathematical function that contains information about the quantum state of a system. In simple terms, it represents the probability amplitude of finding a particle in a particular state at a given position and time. The Schrödinger equation comes in two forms: the time-dependent Schrödinger equation and the time-independent Schrödinger equation. The time-dependent Schrödinger equation describes how the wavefunction changes with time and is given by the equation $i\hbar\partial\Psi/\partial t = H\Psi$, where ih is the imaginary unit multiplied by Planck's constant, $\partial\Psi/\partial t$ represents the partial derivative of Ψ with respect to time, and H is the Hamiltonian operator. The Hamiltonian operator represents the total energy of the quantum system and includes both the kinetic energy and the potential energy of the particles in the system. In essence, this equation describes how the quantum state evolves in time under the influence of the system's *energy. The time-independent Schrödinger equation, on the other hand, is used to find the allowed energy levels, or eigenvalues, of a quantum system. It is expressed as $H\Psi = E\Psi$, where H is the Hamiltonian operator, Ψ is the wavefunction, and E represents the energy eigenvalue. Solving this equation allows us to determine the quantized energy levels that a particle in a quantum system can have. These energy levels correspond to the discrete energy states of the system. The Schrödinger equation is a foundational concept in quantum mechanics and is essential for understanding the behavior of electrons, atoms, molecules, and other quantum systems. One of the key features of the Schrödinger equation is its probabilistic nature. Unlike classical physics, where the state of a particle can be precisely determined, quantum mechanics introduces an inherent uncertainty. The wavefunction Ψ provides a probabilistic description of the particle's position and properties. The square of the absolute value of the wavefunction, $|\Psi|^2$, represents the probability density, meaning that the likelihood of finding a particle in a specific region of space is proportional to $|\Psi|^2$ in that region. This probabilistic nature of quantum mechanics is encapsulated in the famous statement by Max Born that the square of the wavefunction gives the probability of finding a particle in a particular state. The Schrödinger equation also reveals the wave-particle duality of particles in quantum mechanics. Particles, such as electrons, can exhibit both wave-like and particle-like behavior depending on the circumstances. When not observed, particles can exist in a* *superposition of states, represented by the wavefunction, allowing them to exhibit interference patterns and diffraction, similar to waves. However, upon measurement, the particle's state collapses to a single outcome, behaving like a classical particle. This dual nature of particles is a fundamental aspect of quantum mechanics and is embedded in the mathematical framework of the Schrödinger equation. Solving the Schrödinger equation for complex quantum systems can be a highly challenging task. Exact solutions are only available for a limited set of simple systems, such as the hydrogen atom. For more complex systems, numerical and computational methods are often employed to approximate the wavefunction and calculate the energy levels and properties of the quantum system. These methods involve discretizing space and time and using iterative techniques to approach the solutions. In practice, solving the Schrödinger equation for molecules and solids is a cornerstone of computational chemistry and materials science, enabling the prediction of molecular structures, reaction mechanisms, and material properties. Furthermore, the Schrödinger equation plays a vital role in understanding chemical bonding and the behavior of electrons in atoms and molecules. It forms the basis for molecular orbital theory, which describes how electrons are distributed among molecular orbitals in molecules. Molecular orbitals are solutions to the Schrödinger equation that describe the probability distribution of electrons in a molecule. The combination of molecular orbitals gives rise to bonding and antibonding orbitals, which determine the stability and properties of molecules. The Schrödinger equation has also been successfully applied to understand the behavior of particles in electromagnetic fields. In quantum electrodynamics (QED), the theory that describes the interaction of electrons and photons, the Schrödinger equation is extended to incorporate the effects of relativistic quantum mechanics and quantum field theory. This allows for the accurate prediction of electromagnetic interactions at the quantum level, including phenomena like electron-photon scattering and the Lamb shift. In the realm of quantum mechanics, the Schrödinger equation is a foundational equation that has profound implications for our understanding of the quantum world. It provides a mathematical framework for describing the behavior of particles in quantum systems, from electrons in atoms to the behavior of molecules and materials. The probabilistic and wave-like nature of quantum systems, as well as the wave-particle duality of particles, are inherent features revealed by the Schrödinger equation. Solving this equation is a central task in quantum mechanics, allowing us to calculate the quantized energy levels of systems and understand their properties. In the modern era of computational science, the Schrödinger equation is a cornerstone of research in fields ranging from chemistry and materials science to quantum electrodynamics and quantum computing. It continues to shape our understanding of the quantum world and drive innovations in science and technology. The Schrödinger equation serves as the foundational* *equation of quantum mechanics, providing a powerful tool for solving a wide range of quantum problems. It allows us to understand the behavior of quantum systems by describing the evolution of their wavefunctions in both time-dependent and time-independent contexts. When applied to specific physical systems, the Schrödinger equation yields valuable insights into the properties and behavior of particles at the quantum level. Solving quantum problems with the Schrödinger equation typically begins with defining the Hamiltonian operator, denoted as H. The Hamiltonian operator represents the total energy of the quantum system and includes terms for kinetic energy and potential energy. By constructing the appropriate Hamiltonian for a given system, we can formulate the Schrödinger equation that governs its behavior. In the time-dependent Schrödinger equation, $i\hbar \partial \Psi / \partial t = H\Psi$, the wavefunction Ψ evolves with time under the influence of the Hamiltonian operator H. This equation describes how the quantum state of the system changes dynamically over time, taking into account the system's energy. Solving the time-dependent Schrödinger equation yields the time evolution of the wavefunction, providing information about how the quantum state of the system evolves. In practical applications, the time-dependent Schrödinger equation is used to model dynamic quantum processes, such as chemical reactions, electron dynamics, and quantum transport. Solving this equation allows us to predict how a quantum system will change over time, making it a crucial tool in understanding real-world phenomena. In* *contrast, the time-independent Schrödinger equation, HΨ = EΨ, is used to find the allowed energy levels, or eigenvalues, of a quantum system. This equation focuses on the stationary states of the system, where the wavefunction does not change with time. The solutions to the time-independent Schrödinger equation provide information about the quantized energy levels that the quantum system can occupy. These energy levels correspond to the discrete states of the system and are associated with specific quantum properties. To solve the time-independent Schrödinger equation, one typically employs mathematical techniques and approximations. Exact solutions are available for a limited set of simple systems, such as the hydrogen atom, where the Schrödinger equation can be solved analytically. However, for more complex systems, numerical methods and computational approaches are often necessary. One widely used technique is the variational method, where trial wavefunctions are proposed, and the energy is minimized by adjusting parameters to approximate the true wavefunction. Numerical methods, such as the Hartree-Fock method and density functional theory, are employed to tackle complex quantum systems, including molecules and materials. These methods use approximations to efficiently solve the Schrödinger equation and predict properties like molecular structures and electronic configurations. The Schrödinger equation is a cornerstone of computational chemistry and materials science, enabling the study of diverse systems. It plays a crucial role in understanding chemical bonding, electronic structure, and the behavior of electrons in atoms and molecules. By solving the time-independent Schrödinger equation, researchers can determine the electronic energy levels and molecular properties that govern chemical reactions and material properties. Furthermore, the Schrödinger equation has applications in the field of quantum optics, where it describes the behavior of photons and their interactions with matter. Quantum optics explores phenomena like quantum interference, photon absorption, and emission, all of which are mathematically described by the Schrödinger equation. In quantum electrodynamics (QED), an extension of the Schrödinger equation incorporates relativistic quantum mechanics and quantum field theory to explain the interaction of charged particles with electromagnetic fields. This theory is instrumental in understanding phenomena like electron-photon scattering and the Lamb shift. Quantum mechanics also gives rise to quantum computing, a field that exploits the principles of superposition and entanglement. The Schrödinger equation forms the basis for quantum algorithms, which are algorithms designed to be executed on quantum computers. These algorithms take advantage of the quantum properties of qubits, such as superposition and entanglement, to perform certain calculations exponentially faster than classical computers. Notable quantum algorithms include Shor's algorithm, which can efficiently factor large numbers, and Grover's algorithm, which searches unsorted databases more quickly than classical algorithms. As quantum computing technology advances, solving complex problems that were once infeasible becomes a reality. In addition to its foundational role in quantum mechanics, the Schrödinger equation has profound philosophical implications for our understanding of the nature of reality and the role of observation. The probabilistic nature of quantum mechanics, as described by the Schrödinger equation, challenges classical determinism. Particles exist in superpositions of states until measured, at which point their wavefunctions collapse to a single outcome. This raises questions about the fundamental nature of reality and the role of observation in quantum measurements. The famous double-slit experiment, where particles exhibit both wave-like and particle-like behavior, underscores the uncertainty inherent in quantum mechanics. The Schrödinger equation reveals that our knowledge of a quantum system is probabilistic, and the act of measurement can fundamentally alter its state. This aspect of quantum mechanics continues to spark debates and discussions among physicists and philosophers. In summary, the Schrödinger equation is a powerful tool for solving a wide range of quantum problems and understanding the behavior of quantum systems. Whether applied to dynamic processes through the time-dependent Schrödinger equation or used to determine energy levels and properties through the time-independent Schrödinger equation, it is an essential component of quantum mechanics. Solving the Schrödinger equation allows us to predict the behavior of quantum systems, from chemical reactions and electronic structure to quantum computing and quantum optics. Furthermore, the probabilistic nature of quantum mechanics, as described by the Schrödinger equation, challenges our classical intuitions about reality and observation, opening doors to profound philosophical discussions about the nature of the quantum world.*

Chapter 8: Quantum States and Operators: Building Blocks of Reality

In quantum mechanics, the description of a quantum system's state is typically represented by a mathematical entity known as a quantum state vector. This vector encapsulates all the information about the quantum system and is essential for predicting and understanding its behavior. Quantum state vectors provide a complete description of a quantum system's properties, including its position, momentum, energy, and other observable quantities. The quantum state vector, often denoted as $|\Psi\rangle$ (the "ket" notation), is an abstract mathematical object that exists in a complex vector space called a Hilbert space. The Hilbert space associated with a quantum system depends on the specific properties of that system. For example, the Hilbert space for a single particle in three-dimensional space is different from the Hilbert space for a system of multiple particles with spin. In quantum mechanics, each quantum state vector $|\Psi\rangle$ corresponds to a unique quantum state of the system. These quantum states can represent the position, momentum, or any other observable property of the system. One of the key principles of quantum mechanics is that the square of the magnitude of the quantum state vector $|\Psi\rangle$, denoted as $|\Psi|^2$, represents the probability density of finding the system in a particular state. This means that $|\Psi|^2$ gives the likelihood of measuring a specific value *for a given observable. The probability density is always a real and non-negative quantity, ensuring that the probabilities are physically meaningful. Quantum state vectors are used to describe both discrete and continuous systems. For discrete systems, such as a quantum spin system with two possible states (spin-up and spin-down), the quantum state vector can be represented as a linear combination of basis states. In this case, the quantum state vector $|\Psi\rangle$ can be written as $|\Psi\rangle = \alpha|\uparrow\rangle + \beta|\downarrow\rangle$, where α and β are complex coefficients, and $|\uparrow\rangle$ and $|\downarrow\rangle$ are the basis states corresponding to spin-up and spin-down, respectively. The coefficients α and β are usually complex numbers because quantum mechanics allows for complex probability amplitudes. To obtain the probabilities of measuring a specific outcome, one would calculate the squared magnitude of the coefficients $|\alpha|^2$ and $|\beta|^2$. For continuous systems, such as the position of a particle, the quantum state vector $|\Psi\rangle$ can be represented as a wavefunction. The wavefunction, often denoted as $\Psi(x)$, describes the probability amplitude of finding the particle at a particular position x. In this context, the wavefunction $\Psi(x)$ is a complex-valued function of position. The square of the magnitude of the wavefunction $|\Psi(x)|^2$ represents the probability density of finding the particle at position x. The wavefunction $\Psi(x)$ is a fundamental concept in quantum mechanics and plays a central role in understanding the behavior of quantum systems. The Schrödinger equation, a foundational equation in quantum mechanics, describes how the wavefunction $\Psi(x)$ evolves over time for a given* *quantum system. This evolution is governed by the Hamiltonian operator, which represents the total energy of the system. Solving the Schrödinger equation allows us to determine how the wavefunction changes with time and predict the quantum state of the system at any future time. Wavefunctions provide valuable insights into the behavior of quantum systems. For example, the shape of the wavefunction can reveal information about the spatial distribution of a particle. A wavefunction that is more concentrated in a particular region indicates a higher probability of finding the particle there. Conversely, a wavefunction that is spread out over a larger region implies a lower probability of finding the particle in any specific location within that region. Additionally, wavefunctions are used to calculate expected values of observables, such as position and momentum. The expectation value of an observable A, denoted as $\langle A \rangle$, is obtained by integrating the product of the wavefunction $\Psi(x)$ and the operator corresponding to the observable. For example, the expectation value of position $\langle x \rangle$ is calculated as $\langle x \rangle = \int x |\Psi(x)|^2 \, dx$, where the integral extends over all possible positions. Similarly, the expectation value of momentum $\langle p \rangle$ is determined by integrating $p |\Psi(x)|^2$ over all possible momenta. Wavefunctions can also represent superposition states, where a quantum system is in a linear combination of multiple states. A classic example of superposition is the double-slit experiment, where particles exhibit both wave-like and particle-like behavior. In this experiment, particles are in a superposition of states when they pass through two slits, creating an interference pattern on a* *screen. The wavefunction describes the probability amplitudes of the different states and their interference effects. Superposition states are a fundamental aspect of quantum mechanics and are essential for understanding phenomena like quantum entanglement. Quantum entanglement occurs when two or more particles are in a superposition of states that are correlated with each other. The entangled particles remain connected, even when separated by large distances, and measuring one particle's state instantaneously determines the state of the other(s). This non-local correlation, often referred to as "spooky action at a distance," is described mathematically by the entangled state's wavefunction. In summary, quantum state vectors and wavefunctions are fundamental concepts in quantum mechanics that provide a mathematical description of a quantum system's properties. They allow us to calculate probabilities, predict outcomes of measurements, and understand the behavior of particles in quantum systems. Whether representing discrete systems as quantum state vectors or continuous systems as wavefunctions, these mathematical representations play a central role in our ability to explore and manipulate the quantum world. In the realm of quantum mechanics, operators play a fundamental and central role. Operators are mathematical entities that represent physical observables and transformations in quantum systems. They provide a bridge between the mathematical framework of quantum mechanics and the real-world phenomena that we observe and measure. Operators* *are essential tools for understanding and predicting the behavior of quantum systems. One of the most common and fundamental operators in quantum mechanics is the Hamiltonian operator, denoted as H. The Hamiltonian operator represents the total energy of a quantum system, which includes both kinetic energy and potential energy. In quantum mechanics, the time-dependent Schrödinger equation $i\hbar\partial\Psi/\partial t = H\Psi$ describes how the wavefunction Ψ of a quantum system changes over time under the influence of the Hamiltonian operator H. This equation is a cornerstone of quantum mechanics, as it governs the dynamics of quantum systems and provides a mathematical framework for understanding the evolution of quantum states. Another important operator in quantum mechanics is the position operator, denoted as X. The position operator X corresponds to the physical observable of position, and its eigenvalues represent the possible positions of a particle in a quantum system. The position operator acts on the wavefunction Ψ, and its action results in the expected value or mean position of the particle. The momentum operator, denoted as P, is another fundamental operator in quantum mechanics. The momentum operator corresponds to the physical observable of momentum, and its eigenvalues represent the possible momenta of a particle. The momentum operator is related to the derivative of the position operator with respect to position, and it plays a crucial role in the Heisenberg uncertainty principle, which imposes limits on the precision with which position and momentum can be simultaneously known in a quantum system.*

Operators can also represent angular momentum, spin, and other physical observables. For example, the angular momentum operator L represents the angular momentum of a particle, while the spin operator S represents the intrinsic angular momentum or spin of a particle. These operators have eigenvalues that correspond to quantized angular momentum values, and they play a crucial role in describing the behavior of particles in quantum systems, such as electrons in atoms. The operators discussed so far are known as Hermitian operators, which have the property that their eigenvalues are real, and their eigenstates are orthogonal. Hermitian operators are essential in quantum mechanics because they correspond to physical observables that yield real measurements. Observables such as position, momentum, energy, and angular momentum are all represented by Hermitian operators in quantum mechanics. The mathematical properties of Hermitian operators also ensure that the expectation values of observables are real numbers, which is consistent with the results of measurements. In addition to Hermitian operators, there are non-Hermitian operators in quantum mechanics, which do not have the property of real eigenvalues. Non-Hermitian operators are often used in the study of open quantum systems and quantum decay processes. These operators describe the coupling of a quantum system to its environment and can account for phenomena like quantum dissipation and decoherence. Operators can be combined through mathematical operations to create new operators. For example, the Hamiltonian operator

H, representing the total energy of a system, is often expressed as the sum of kinetic energy and potential energy operators. H = T + V, where T is the kinetic energy operator and V is the potential energy operator. This decomposition allows us to study the individual contributions of kinetic and potential energy to the behavior of a quantum system. Operators can also be used to define physical transformations or unitary operations in quantum mechanics. Unitary operators, denoted as U, are operators that preserve the inner product of quantum states, meaning that they do not change the probabilities of measurement outcomes. Unitary operators are essential for describing the time evolution of quantum states and for implementing quantum gates in quantum computing. Quantum gates are unitary operations that manipulate the quantum state of a qubit or quantum register. They are the building blocks of quantum algorithms and quantum circuits. The role of operators in quantum mechanics extends beyond the description of physical observables and transformations. Operators also play a crucial role in defining the quantum state of a system. A quantum state vector $|\Psi\rangle$ can be represented as a linear combination of basis states, each of which corresponds to an eigenstate of a particular operator. For example, in the position basis, the quantum state vector $|\Psi\rangle$ can be represented as $|\Psi\rangle = \int \psi(x)|x\rangle\,dx$, where $\psi(x)$ is the wavefunction in the position basis, $|x\rangle$ represents position eigenstates, and the integral extends over all possible positions. Operators are used to switch between different bases and to express quantum states *in the basis that is most convenient for a given problem. The inner product of quantum states, also known as the bra-ket notation, is another important concept involving operators. The inner product of two quantum states, represented as $\langle\Phi|\Psi\rangle$, quantifies the degree of overlap or similarity between the states. It is calculated by applying the Hermitian conjugate of one state to the other and then taking their inner product. The inner product is used to calculate probabilities and to determine the amplitude of transitioning from one state to another. Operators in quantum mechanics are not limited to one-dimensional or single-particle systems. They can also be applied to multi-particle systems, composite systems, and entangled states. In these cases, operators act on the combined Hilbert space of all particles involved, and their eigenvalues and eigenstates describe the collective properties of the system. Operators provide a powerful mathematical framework for understanding the behavior of quantum systems and for making predictions about measurement outcomes. They allow us to represent physical observables, describe transformations, define quantum states, and analyze the properties of quantum systems. In essence, operators serve as the language of quantum mechanics, enabling us to translate the abstract mathematics of quantum theory into concrete predictions and observations in the physical world.*

Chapter 9: Quantum Experiments: Peering into the
Subatomic Realm

The double-slit experiment is one of the most iconic and thought-provoking experiments in the history of physics. It showcases the fascinating phenomenon known as wave-particle duality, which challenges our classical intuition about the nature of particles. The experiment was first conducted by Thomas Young in 1801 and has since been refined and extended to various particles, from electrons to photons.

At its core, the double-slit experiment involves shining a beam of particles, such as electrons or photons, at a barrier with two slits. Behind the barrier, there is a screen on which the particles can be detected. In a classical scenario, one would expect the particles to behave like tiny bullets and create two distinct bands on the screen, aligned with the two slits. However, the reality is far more intriguing. When particles are sent through the double slits one at a time, something remarkable happens. Instead of producing two simple bands, they create an interference pattern on the screen, akin to the pattern produced by waves. This interference pattern consists of alternating light and dark bands, much like the ripples on the surface of a pond when two sets of waves meet. This baffling behavior challenges our classical understanding because particles, such as electrons, were traditionally thought to be discrete, localized entities, quite different from the continuous, wavelike behavior exhibited by light and other wave phenomena. The interference pattern observed in the double-slit experiment implies that particles possess wave-like properties. In other words, they exhibit both particle-like and wave-like characteristics simultaneously. This peculiar phenomenon is encapsulated in the concept of wave-particle duality. Wave-particle duality suggests that particles, at the quantum level, do not have definite positions or trajectories until they are measured. Instead, they exist as probability waves, described by wavefunctions, which represent the likelihood of finding a particle at different positions. The interference pattern arises because the wavefunctions from the two slits overlap and interfere with each other. Where the waves align, they reinforce, creating bright bands, and where they cancel out, they produce dark bands. This behavior is a fundamental aspect of quantum mechanics, and it challenges our classical intuition.

The double-slit experiment reveals that the act of measurement collapses the probability wave and forces the particle to assume a definite position. In other words, particles transition from a state of superposition, where they exist in multiple possible states simultaneously, to a single, well-defined state upon measurement. This transition is often referred to as wavefunction collapse. The remarkable aspect of the double-slit experiment is that even when particles are sent through the slits one at a time, and there is no interference between them, the interference pattern still emerges over time as more and more particles accumulate on the screen. This phenomenon raises profound questions about the nature of reality and the role of observation in quantum mechanics. It suggests that particles somehow "know" about the presence of the other slit and adjust their behavior accordingly, even when no other particles are present to interact with. This mysterious behavior challenges the classical concept of causality, where an effect is determined by a preceding cause. In the quantum realm, it appears that particles can be influenced by future measurements, suggesting a deep and enigmatic connection between the quantum world and the act of observation.

The double-slit experiment has been conducted with various types of particles, and the results remain consistent across all experiments. For instance, when electrons are used in the experiment, they exhibit wave-particle duality, producing interference patterns. Similarly, photons, the particles of light, also display this behavior. The implications of wave-particle duality extend beyond the double-slit experiment and have profound consequences for our understanding of quantum mechanics. One of the key aspects of quantum mechanics is that particles can exist in superposition states, where they simultaneously occupy multiple states or positions until they are measured. This feature is essential for quantum computing and quantum cryptography, where quantum bits, or qubits, can exist as combinations of 0 and 1, allowing for exponentially *faster computation and secure communication. Wave-particle duality also plays a critical role in quantum entanglement, a phenomenon where the properties of two or more particles become correlated in such a way that measuring one particle instantly determines the state of the others, regardless of the distance between them. Albert Einstein famously referred to this phenomenon as "spooky action at a distance." The concept of wave-particle duality has inspired ongoing debates and discussions about the fundamental nature of particles and the philosophical implications of quantum mechanics. It challenges our classical intuitions and forces us to grapple with the idea that the quantum world operates according to rules that defy our everyday experiences.*

Some interpretations of quantum mechanics, such as the Copenhagen interpretation, suggest that the act of measurement is the ultimate arbiter of a particle's behavior, and the wavefunction collapse is a fundamental aspect of the quantum world. Others, like the Many-Worlds interpretation, propose that all possible outcomes occur in separate, parallel realities, preserving the superposition of particles. Regardless of the interpretation one subscribes to, the double-slit experiment and wave-particle duality underscore the richness and complexity of the quantum world. They remind us that our classical intuitions may not always apply in the realm of quantum mechanics and encourage us to explore the profound mysteries of the quantum universe further.

Quantum experiments in particle physics have played a pivotal role in advancing our understanding of the fundamental building blocks of the universe. These experiments delve into the subatomic realm, probing particles at energies and scales that are far beyond our everyday experiences. Particle physics, also known as high-energy physics, seeks to unravel the mysteries of the quantum world and the fundamental forces that govern it. One of the primary goals of particle physics is to discover new particles and elucidate their properties. Quantum experiments in this field often involve particle accelerators, massive machines that accelerate particles to nearly the speed of light. These accelerated particles are then collided or directed into targets, producing high-energy interactions that yield valuable insights into the subatomic world.

One of the most famous particle accelerators is the Large Hadron Collider (LHC) at CERN in Geneva, Switzerland. The LHC is a behemoth, with a circumference of approximately 17 miles, and it smashes protons at unprecedented energies. Quantum experiments at the LHC have led to remarkable discoveries, including the observation of the Higgs boson, a particle responsible for imparting mass to other particles. The existence of the Higgs boson was predicted by the Standard Model of particle physics, a theory that describes the fundamental particles and their interactions. The confirmation of the Higgs boson's existence in 2012 was a monumental achievement and a testament to the power of quantum experiments in particle physics. These experiments involved analyzing the debris from high-energy proton collisions to identify the telltale signs of the Higgs boson's presence. Quantum experiments in particle physics also explore the nature of forces and particles beyond those described by the Standard Model. One of the mysteries that these experiments aim to unravel is the nature of dark matter, a form of matter that does not emit or interact with light but makes up a significant portion of the universe's mass. Although dark matter's existence is inferred from its gravitational effects, its exact composition remains unknown.

Quantum experiments involving sensitive detectors and deep underground laboratories are ongoing in the quest to detect and identify dark matter particles. Another intriguing aspect of particle physics is the study of neutrinos, tiny and elusive particles that barely interact with other matter. Neutrinos come in three flavors: electron, muon, and tau neutrinos. Quantum experiments have revealed that neutrinos can undergo a phenomenon known as neutrino oscillation, where one flavor of neutrino can transform into another as they travel through space. This discovery implies that neutrinos have mass, challenging earlier assumptions. The search for the absolute masses of neutrinos and the determination of their properties remain active areas of research in particle physics. Quantum experiments have also played a crucial role in elucidating the strong nuclear force, one of the fundamental forces that bind quarks together to form protons, neutrons, and other hadrons. Quantum chromodynamics (QCD) is the theory that describes the strong force and the behavior of quarks and gluons, the particles that mediate the force. Quantum experiments, such as deep inelastic scattering and studies of the strong force at high energies, have provided key insights into the nature of QCD and the behavior of quarks and gluons within hadrons. Quantum experiments in particle physics have not only expanded our understanding of the subatomic world but have also pushed the boundaries of technology and innovation.

Accelerators and detectors developed for these experiments have found applications in fields beyond particle physics, including medical imaging, materials science, and industry. The development of particle detectors capable of capturing and analyzing the incredibly high-energy collisions produced by particle accelerators has led to advancements in sensor technology and imaging techniques. Quantum experiments have also contributed to our understanding of the early universe. Through the study of cosmic microwave background radiation and the abundance of light elements, particle physicists have helped to develop the framework for our current understanding of the Big Bang theory. These experiments have provided crucial evidence supporting the idea that the universe began as a hot, dense state and has been expanding and cooling ever since. Additionally, quantum experiments have explored the properties of particles and forces that existed in the early moments of the universe, shedding light on the conditions that prevailed *during the first moments after the Big Bang. The exploration of the quantum world in particle physics is not limited to the study of particles but also extends to the investigation of fundamental symmetries. Symmetries play a fundamental role in the laws of physics and are intimately connected to the conservation laws that govern physical processes. For example, the principle of conservation of angular momentum arises from the rotational symmetry of physical laws. Quantum experiments have tested various symmetries, including time-reversal symmetry and parity symmetry, and have uncovered instances where these symmetries are violated at the quantum level. These violations provide valuable clues about the fundamental forces and interactions in the universe.*

Quantum experiments have also probed the phenomenon of CP violation, which involves the violation of the combined symmetry of charge conjugation (C) and parity (P). CP violation is essential for explaining the predominance of matter over antimatter in the universe, a puzzle known as the matter-antimatter asymmetry. Understanding the source of CP violation is a major goal of particle physics research. In summary, quantum experiments in particle physics have been instrumental in advancing our understanding of the quantum world, uncovering new particles, probing fundamental forces, and testing symmetries. These experiments have not only expanded our knowledge of the subatomic realm but have also led to technological innovations with applications beyond physics. Furthermore, they have contributed to our understanding of the early universe and the fundamental principles that govern the cosmos. As particle accelerators continue to push the boundaries of energy and precision, the field of particle physics remains at the forefront of scientific exploration, promising even more exciting discoveries in the future.

Chapter 10: Quantum Applications: From Quantum Computing to Quantum Teleportation

Quantum computing represents a revolutionary paradigm in the world of information processing. Unlike classical computers that rely on bits as the smallest unit of data, quantum computers employ quantum bits or qubits. The unique property of qubits is their ability to exist in a superposition of states, allowing them to represent both 0 and 1 simultaneously. This inherent duality enables quantum computers to explore an exponential number of possibilities in parallel. Quantum computing has the potential to solve complex problems that are currently intractable for classical computers. One of the most well-known quantum algorithms is Shor's algorithm, which can efficiently factor large numbers. This has significant implications for cryptography, as many encryption methods rely on the difficulty of factoring large numbers. Quantum computers could potentially break these encryption schemes, necessitating the development of quantum-resistant cryptographic techniques. Another notable quantum algorithm is Grover's algorithm, which can search an unsorted database of N items in O(sqrt(N)) steps, offering a quadratic speedup over classical algorithms. This has applications in data search and optimization problems. Quantum computing also promises advances in drug discovery and materials science by simulating complex quantum systems with high accuracy. Such simulations can lead to the discovery of new drugs and materials with unprecedented efficiency. Additionally, quantum computing has the potential to revolutionize machine learning and artificial intelligence by accelerating tasks like optimization, pattern recognition, and natural language processing. Quantum machine learning algorithms can exploit quantum parallelism to provide significant speedups. Quantum computing's power is not solely theoretical; several companies and research institutions have built and tested quantum computers. Quantum computers vary in size and capabilities, from small-scale quantum processors with a few qubits to more advanced devices with dozens or even hundreds of qubits. Companies like IBM, Google, and Rigetti are actively developing quantum hardware and making it accessible through cloud-based platforms. These platforms allow researchers and developers to experiment with quantum algorithms and applications. Despite these advancements, quantum computing faces significant challenges and limitations. One major obstacle is qubit coherence time, which refers to the duration qubits can maintain their superposition and entanglement. Currently, coherence times are relatively short, making it challenging to perform complex computations. Quantum error correction codes have been proposed to address this issue, but implementing them is a formidable task. Another limitation is qubit connectivity, as qubits need to be interconnected to perform operations. Achieving high connectivity among qubits in a scalable way remains a technological *challenge. Furthermore, building fault-tolerant quantum computers that can reliably execute quantum algorithms is a goal that researchers are actively pursuing. Quantum computers operate at extremely low temperatures to reduce environmental noise and maintain qubit coherence. This necessitates specialized cooling systems, making quantum computing expensive and cumbersome. To make quantum computing more accessible and practical, researchers are exploring quantum annealing, a different approach that focuses on optimization problems. Companies like D-Wave Systems are developing quantum annealers that are specialized for specific tasks, such as optimization and machine learning. These devices are commercially available and used in various applications. Quantum computing has the potential to revolutionize industries and solve problems that were previously insurmountable. It holds promise in fields as diverse as cryptography, drug discovery, materials science, machine learning, and more. However, the field is still in its early stages, and many technical challenges must be overcome to realize its full potential. As quantum computers continue to evolve and mature, they are likely to play an increasingly prominent role in our computational landscape. Quantum algorithms that can efficiently solve problems with real-world implications are actively being researched. The field of quantum computing is highly interdisciplinary, drawing from quantum physics, computer science, mathematics, and engineering. Researchers are continually pushing the boundaries of what quantum computers can achieve,* exploring new algorithms and applications. Quantum supremacy, a milestone where quantum computers outperform the best classical computers for a specific task, has been achieved for certain problems. However, it remains to be seen how quantum computing will impact society on a broader scale. One of the challenges is integrating quantum algorithms into existing computational workflows. Hybrid quantum-classical approaches are being developed to combine the strengths of both classical and quantum computing. These approaches leverage quantum computers for specific tasks while utilizing classical computers for others. Quantum computing is not a replacement for classical computing but rather a complementary technology. The advent of quantum computing raises questions about the security of classical encryption methods. As quantum computers advance, they pose a potential threat to the security of data encrypted with currently used methods. To address this, researchers are exploring post-quantum cryptography, which aims to develop encryption techniques that are resistant to quantum attacks. The transition to quantum-resistant encryption will be a crucial aspect of securing data in a post-quantum world. Quantum computing also has implications for information theory and the study of the fundamental limits of computation. It challenges our understanding of computation and what can be achieved with the laws of quantum physics. The development of quantum algorithms has led to new insights into the relationships between classical and quantum information theory. Quantum computing *research is a global endeavor, with contributions from academic institutions, industry leaders, and startups worldwide. Collaborations between researchers and organizations are essential for advancing the field and addressing its challenges. Quantum computing has the potential to transform industries and revolutionize problem-solving across various domains. As quantum hardware and algorithms continue to progress, their impact on society will become increasingly profound. While the path to practical and scalable quantum computing is not without obstacles, the potential rewards are significant. Quantum computing represents a new frontier in technology and science, offering the promise of solving complex problems that were once thought to be beyond reach. Quantum teleportation is a fascinating phenomenon in the realm of quantum mechanics. It is not the teleportation of physical objects but rather the transmission of quantum information from one location to another. This process involves the entanglement of two particles, known as entangled qubits, and the transfer of quantum information from one qubit to another through measurement and classical communication. The concept of quantum teleportation was first proposed by Charles H. Bennett, Gilles Brassard, Claude Crépeau, Richard Jozsa, Asher Peres, and William K. Wootters in 1993. Quantum teleportation exploits the principles of quantum entanglement, superposition, and quantum measurement to achieve a seemingly instantaneous transfer of quantum states. The fundamental idea behind quantum teleportation is to use two entangled qubits as a quantum channel to transmit information about the quantum state of a third qubit. The process begins with the entanglement of two qubits, typically referred to as qubit A and qubit B. These entangled qubits can be created through various methods, such as entanglement swapping or entangling two particles initially in a superposition of states. Once qubits A and B are entangled, they share a special quantum connection that allows information to be transmitted between them instantaneously. Now, consider a third qubit, qubit C, that holds the quantum state to be teleported. The goal is to transfer the quantum state from qubit C to qubit B, which may be located far away from each other. The teleportation process unfolds in several steps, with the assistance of quantum gates and measurements. First, qubits A and C undergo a specific quantum operation called a Bell measurement. This measurement entangles qubits A and C in such a way that the measurement result on qubit A conveys information about the state of qubit C. After the Bell measurement on qubits A and C, a classical message is sent to the location of qubit B, which contains the outcome of the measurement. This message includes two classical bits of information. With this information, the receiver at qubit B performs a set of quantum operations on qubit B, conditioned on the measurement results received. These operations effectively transform qubit B into an exact replica of the initial quantum state of qubit C. The quantum state from qubit C has now been teleported to qubit B, despite the physical separation between them. It's important to* note that quantum teleportation doesn't involve the instantaneous movement of particles from one location to another, as often depicted in science fiction. Instead, it relies on the principles of quantum entanglement and measurement to transmit information about the quantum state of a particle. Quantum teleportation has been experimentally demonstrated and is a powerful tool in the field of quantum information processing. It forms the basis for various quantum communication protocols and quantum computing algorithms. One of the significant applications of quantum teleportation is in quantum cryptography, where it can be used to securely transmit quantum encryption keys. Quantum key distribution (QKD) protocols, such as the well-known BB84 protocol, use quantum teleportation to establish secure communication channels. By transmitting quantum encryption keys using teleportation, it becomes exceedingly difficult for eavesdroppers to intercept or tamper with the keys, as any measurement or disturbance would be detectable. Quantum teleportation also plays a crucial role in quantum computing, especially in the context of quantum error correction. Quantum error correction codes are essential for mitigating the effects of noise and decoherence that naturally occur in quantum computers. Teleportation-based error correction protocols allow for the detection and correction of errors in quantum computations, making quantum algorithms more reliable and accurate. Furthermore, quantum teleportation has applications in quantum networking, enabling the distribution of entangled qubits across long distances. This is a fundamental building block for the development of quantum networks that can connect quantum computers and secure communication nodes. Quantum teleportation also holds potential in the emerging field of quantum internet, where quantum information can be transmitted over global-scale quantum networks. The ability to teleport quantum states across vast distances could revolutionize secure communication, distributed quantum computing, and quantum-enhanced sensing. Quantum teleportation experiments have pushed the boundaries of our understanding of quantum mechanics and have demonstrated the remarkable potential of quantum information processing. These experiments have achieved teleportation over increasingly longer distances, including from the ground to a satellite in space. The field of quantum teleportation continues to advance, with ongoing research aimed at improving the efficiency and fidelity of the teleportation process. While quantum teleportation is a powerful tool for transmitting quantum information, it is not without its challenges. One significant challenge is the need for a pre-established entangled state between qubits A and B, which requires careful preparation and maintenance. Creating and preserving entanglement over long distances or in noisy environments can be technically demanding. Additionally, the process of quantum measurement and classical communication introduces the need for classical information transfer, which limits the speed of quantum teleportation to the speed of light. Quantum teleportation is, therefore, not a means of achieving faster-than-light communication, as it relies *on classical channels for information exchange. Despite these challenges, quantum teleportation remains a cornerstone of quantum information science and has opened new avenues for secure communication, quantum computing, and quantum networking. As the field of quantum technology continues to advance, the practical applications of quantum teleportation are expected to grow, contributing to the development of a quantum-enhanced future.*

BOOK 2
FROM STRING THEORY TO QUANTUM COMPUTING
A JOURNEY THROUGH QUANTUM PHYSICS

ROB BOTWRIGHT

Chapter 1: Quantum Physics Primer

The historical development of quantum physics is a journey through scientific breakthroughs and paradigm shifts that have fundamentally altered our understanding of the physical world. It all began in the late 19th century when classical physics, based on the principles of Isaac Newton, seemed to provide a comprehensive description of the universe. At that time, scientists believed that the laws of classical mechanics could explain the behavior of particles, from the motion of planets to the behavior of atoms. However, cracks in this classical framework started to appear as physicists delved deeper into the mysteries of the atomic and subatomic realms. One of the first challenges to classical physics came from the study of black-body radiation, the electromagnetic radiation emitted by a heated object. According to classical theory, the intensity of this radiation should increase without limit as the temperature rises. This prediction, known as the ultraviolet catastrophe, contradicted experimental observations, which showed that the radiation intensity leveled off at high temperatures. In 1900, Max Planck proposed a groundbreaking solution to this problem by introducing the concept of quantization. Planck postulated that the energy of electromagnetic radiation could only take discrete, quantized values, rather than being continuous. This idea led to the development of quantum mechanics, as Planck's quantization concept laid the foundation for a new way of understanding the behavior of particles at the atomic and subatomic levels. Albert Einstein further advanced the quantum revolution with his work on the photoelectric effect in 1905. Einstein proposed that light consists of discrete packets of energy, which he called photons. This idea challenged the classical view of light as a continuous wave and provided compelling evidence for the existence of quantization in nature. Einstein's work earned him the Nobel Prize in Physics in 1921. The birth of quantum mechanics, as we know it today, can be attributed to the contributions of several prominent physicists, including Niels Bohr, Werner Heisenberg, Max Born, and Erwin Schrödinger. In 1913, Niels Bohr introduced the Bohr model of the atom, which successfully explained the spectral lines of hydrogen. Bohr postulated that electrons in an atom can only occupy specific energy levels, and transitions between these levels result in the emission or absorption of quantized energy in the form of photons. This model provided a framework for understanding the discrete nature of atomic spectra. However, the Bohr model had limitations and couldn't explain the behavior of more complex atoms or molecules. In 1925, Werner Heisenberg formulated matrix mechanics, a mathematical framework for quantum mechanics. Heisenberg's uncertainty principle, proposed in 1927, revolutionized our understanding of particles at the quantum level. The principle asserts that certain pairs of properties, such as a particle's position and momentum, cannot be simultaneously known with arbitrary precision. This fundamental uncertainty challenges the determinism of classical physics and highlights the probabilistic nature of quantum mechanics. Meanwhile, Erwin Schrödinger developed wave mechanics, an alternative formulation of quantum mechanics based on wave functions. Schrödinger's equation, known as the Schrödinger equation, describes how the wave function of a quantum system evolves over time. This equation has been a cornerstone of quantum mechanics and is central to understanding the behavior of particles in quantum systems. The development of quantum mechanics led to a more complete and unified understanding of the physical world, with wave-particle duality as one of its key principles. Wave-particle duality suggests that particles, such as electrons and photons, exhibit both wave-like and particle-like properties. This concept challenges classical notions of particle behavior and underscores the probabilistic nature of quantum physics. In 1926, Max Born introduced the interpretation of the wave function as representing the probability amplitude of finding a particle in a particular state. This interpretation resolved the long-standing question of what the wave function represents in quantum mechanics. With these foundational principles in place, quantum mechanics became the framework for understanding the behavior of particles in the quantum realm. It successfully explained a wide range of phenomena, including the behavior of electrons in atoms, the structure of the periodic table, and the behavior of particles in quantum systems. Quantum mechanics also provided the theoretical foundation for quantum field theory, which describes the behavior of particles and fields in relativistic physics. The development of quantum mechanics and quantum field theory had profound implications for our understanding of the physical world and led to numerous groundbreaking discoveries. In 1928, Paul Dirac formulated quantum mechanics in a way that incorporated special relativity, resulting in a relativistic quantum theory that accurately described the behavior of electrons. Dirac's work introduced the concept of antimatter and predicted the existence of positrons, the antiparticles of electrons. The discovery of positrons in experiments provided strong confirmation of the predictions of quantum theory. Quantum mechanics also played a crucial role in the development of nuclear physics. In the 1930s, quantum mechanics was applied to explain the behavior of protons and neutrons in atomic nuclei, leading to the development of nuclear models. This work laid the foundation for our understanding of nuclear structure and the behavior of matter at extremely high energies. Quantum mechanics also led to the development of quantum electrodynamics (QED), a quantum field theory that describes the interactions of electrons and photons. QED has been extraordinarily successful in explaining the behavior of charged particles and electromagnetic phenomena. One of its remarkable achievements is the precise calculation of the anomalous magnetic moment of the electron, which has been confirmed through experiments with astonishing accuracy. The successful application of quantum mechanics to various fields of *physics cemented its status as one of the most powerful and successful scientific theories in history. Despite its tremendous success, quantum mechanics also gave rise to several puzzling and counterintuitive aspects of the quantum world. One of the most famous examples is the Schrödinger's cat thought experiment, proposed by Erwin Schrödinger in 1935. This paradox involves a cat in a sealed box, where its fate depends on the quantum state of a radioactive atom. According to quantum mechanics, until the box is opened and the cat is observed, it is in a superposition of being both alive and dead. This scenario illustrates the strange and paradoxical nature of quantum superposition and the role of measurement in collapsing quantum states. The interpretation of quantum mechanics has been a subject of ongoing debate and remains a topic of interest among physicists and philosophers. Several interpretations, such as the Copenhagen interpretation, the many-worlds interpretation, and the pilot-wave theory, offer different perspectives on the underlying nature of quantum reality. While these interpretations provide valuable insights, the fundamental principles of quantum mechanics remain consistent across all interpretations, making quantum mechanics one of the most well-tested and successful theories in the history of physics. In the decades that followed the development of quantum mechanics, its applications expanded into areas such as quantum chemistry, solid-state physics, and quantum optics. The advent of quantum computing and quantum information theory in the latter half of the 20th century further extended the reach of quantum mechanics into the realm of information processing and cryptography. Quantum mechanics also paved the way for the development of quantum field theories that describe the behavior of particles and fields in the context of the fundamental forces of nature, such as quantum chromodynamics (QCD) for the strong nuclear force and electroweak theory for the electromagnetic and weak forces. The integration of quantum mechanics with the principles of special relativity in quantum field theories has resulted in a comprehensive framework for understanding the behavior of particles and fields in the universe. In summary, the historical development of quantum physics has been a journey of intellectual exploration and scientific innovation. From its humble beginnings as a solution to the ultraviolet catastrophe and the photoelectric effect, quantum mechanics has evolved into a fundamental theory that underlies our understanding of the behavior of particles, atoms, and the fundamental forces of nature. Its applications have extended far beyond the realm of physics, impacting fields such as chemistry, materials science, and information technology. Despite its counterintuitive and puzzling aspects, quantum mechanics has stood the test of time as one of the most successful and enduring theories in the history of science, reshaping our perception of the quantum world. Quantum mechanics, also known as quantum physics or quantum theory, is a branch of physics that fundamentally changes our understanding of the physical world. At its core, quantum mechanics is a mathematical framework that describes the behavior of* particles, atoms, and molecules at the smallest scales. It was developed in the early 20th century to address the limitations of classical physics, which failed to explain certain phenomena at the atomic and subatomic levels. One of the key concepts of quantum mechanics is quantization, which asserts that certain physical quantities, such as energy, angular momentum, and electric charge, come in discrete, quantized units. This means that these quantities cannot take on any arbitrary value but are restricted to specific, quantized values. Quantization is a departure from classical physics, where such quantities were considered continuous and could vary infinitely. The quantization of energy was first introduced by Max Planck in his explanation of black-body radiation and later extended to other physical phenomena. Planck's quantization concept laid the foundation for quantum mechanics and earned him the Nobel Prize in Physics in 1918. Another fundamental concept in quantum mechanics is wave-particle duality. This concept suggests that particles, such as electrons and photons, exhibit both wave-like and particle-like properties. In classical physics, particles were considered distinct entities with well-defined positions and momenta. However, in the quantum world, particles are described by wave functions, which represent the probability distribution of finding a particle in a particular state. The wave-like nature of particles is evident in phenomena like diffraction and interference, where particles exhibit wave-like behavior. This duality challenged the classical notion of particle behavior and led to the development of wave mechanics, one of the two main formulations of quantum mechanics. The second formulation, known as matrix mechanics, was developed independently by Werner Heisenberg and is based on matrix algebra. Heisenberg's work introduced the concept of quantum operators, which represent physical observables such as position, momentum, and energy. These operators act on wave functions, allowing physicists to calculate the expected outcomes of measurements in quantum systems. One of the most famous principles in quantum mechanics is Heisenberg's uncertainty principle. This principle asserts that certain pairs of properties, such as a particle's position and momentum, cannot be simultaneously known with arbitrary precision. The more accurately one property is measured, the less accurately the other can be known, introducing a fundamental limit to our knowledge of quantum systems. The uncertainty principle highlights the inherent probabilistic nature of quantum mechanics and challenges the determinism of classical physics. Quantum superposition is another essential concept in quantum mechanics. It refers to the ability of particles to exist in multiple states simultaneously until they are measured or observed. In other words, a quantum system can be in a superposition of different states, each with its associated probability amplitude, until a measurement is made, causing the system to collapse into one of those states. Superposition is exemplified by the famous thought experiment of Schrödinger's cat, where a cat inside a sealed box can be considered both alive and dead until the box is opened and the cat's state is observed. This concept has profound implications for the behavior of particles and the processing of quantum information. Entanglement is yet another intriguing phenomenon in quantum mechanics. When two or more particles interact in a way that their quantum states become correlated, they are said to be entangled. Entangled particles, such as entangled qubits in quantum computing, exhibit a strong and often non-classical correlation. This means that the measurement of one entangled particle can instantaneously influence the state of the other, regardless of the physical distance separating them. Albert Einstein famously referred to entanglement as "spooky action at a distance" and questioned its implications for the completeness of quantum mechanics. Entanglement has since been experimentally confirmed and plays a crucial role in quantum communication and quantum computing. Quantum mechanics also introduces the concept of quantum states, which describe the possible configurations of a quantum system. A quantum state can be represented by a wave function, a mathematical object that encodes the probabilities of measuring various properties of the system. The Schrödinger equation, a fundamental equation in quantum mechanics, governs the evolution of quantum states over time. It describes how a quantum system's wave function changes in response to its surroundings and the forces acting upon it. The Schrödinger equation provides a deterministic framework for predicting the future behavior of quantum systems, given their initial states and the forces at play. Quantum mechanics challenges *the classical notion of causality, as it introduces the concept of quantum indeterminacy. In classical physics, it was believed that if one knew the initial conditions of a system and the forces acting on it, one could predict its future state with certainty. In contrast, quantum mechanics only allows predictions of probabilistic outcomes for measurements, introducing an element of inherent randomness into the microscopic world. This indeterminacy is not due to limitations in our knowledge but is a fundamental property of quantum systems. The probabilistic nature of quantum mechanics is captured by the Born rule, which relates the square of the absolute value of the wave function to the probability density of finding a particle in a particular state. The Born rule has been experimentally verified countless times and forms the basis for calculating probabilities in quantum systems. Quantum mechanics also encompasses the concept of quantum entanglement, where the properties of two or more particles become interconnected in such a way that the measurement of one particle instantaneously influences the state of the other(s). This phenomenon, famously described by Einstein as "spooky action at a distance," challenges classical notions of locality and independence. The principles of quantum mechanics have been experimentally confirmed through a multitude of experiments, such as the double-slit experiment, which demonstrates the wave-particle duality of particles. These experiments have consistently shown that quantum mechanics accurately describes the behavior of particles at the quantum level, despite its departure from classical intuitions. Quantum mechanics is not just a theoretical framework; it has practical applications in various fields, including quantum computing, quantum cryptography, and quantum sensing. Quantum computing leverages the principles of superposition and entanglement to perform certain computations exponentially faster than classical computers. Quantum cryptography harnesses the unique properties of quantum states for secure communication, while quantum sensors offer unprecedented precision in measurements. In summary, quantum mechanics is a profound and revolutionary theory that has transformed our understanding of the physical world. It challenges classical notions of determinism, introduces the probabilistic nature of quantum systems, and enables technologies that have the potential to revolutionize computing, communication, and sensing. While its principles may seem counterintuitive and puzzling, quantum mechanics has been consistently validated through experiments, making it one of the most successful and accurate theories in the history of physics.*

Chapter 2: The Road to String Theory

Early attempts at unifying the fundamental forces of nature have been a central pursuit in the history of physics. The quest for a unified theory seeks to reconcile the descriptions of the four known fundamental forces: gravity, electromagnetism, the strong nuclear force, and the weak nuclear force. The earliest attempts at unification were driven by the desire to understand the underlying principles governing the physical universe. One of the first unification attempts came from Sir Isaac Newton, who formulated the law of universal gravitation in the late 17th century. Newton's law described how every mass in the universe attracts every other mass through the force of gravity, and it successfully explained the motion of celestial bodies. However, Newton's law of gravity remained distinct from the other three fundamental forces known at the time. In the 19th century, James Clerk Maxwell made significant strides in the unification of electromagnetism. Maxwell's equations, formulated between 1861 and 1865, described how electric and magnetic fields are interrelated and how they propagate as electromagnetic waves. This unification marked a major breakthrough in the understanding of electromagnetism, unifying electric and magnetic phenomena into a single electromagnetic force. Maxwell's equations also predicted the existence of electromagnetic waves, which were later confirmed experimentally and led to the development of technologies like radio and wireless communication. While the unification of electromagnetism was a significant achievement, the other two fundamental forces, the strong and weak nuclear forces, remained separate from this unified framework. The strong nuclear force, which binds protons and neutrons within atomic nuclei, was not fully understood until the development of quantum chromodynamics (QCD) in the 20th century. QCD described the strong force in terms of quarks and gluons, the fundamental particles that make up protons, neutrons, and other hadrons. The weak nuclear force, responsible for processes like beta decay in atomic nuclei, was also described in the mid-20th century through the electroweak theory, which unified the weak nuclear force with electromagnetism. The electroweak theory, developed by Sheldon Glashow, Abdus Salam, and Steven Weinberg, showed that the electromagnetic and weak nuclear forces are two aspects of a single electroweak force at high energies. This successful unification earned the three physicists the Nobel Prize in Physics in 1979. Despite these achievements, the ultimate goal of a fully unified theory of all four fundamental forces remained elusive. Albert Einstein's theory of general relativity, published in 1915, described gravity as the curvature of spacetime caused by mass and energy. General relativity was a profound departure from Newton's theory of gravity, and it successfully explained various gravitational phenomena, including the bending of light by massive objects and the prediction of black holes. Einstein's theory of general relativity represented a significant unification of gravity and spacetime, but it remained separate from the other fundamental forces. Efforts to unify gravity with the other forces of nature led to the development of grand unified theories (GUTs) in the 20th century. GUTs aimed to merge the strong nuclear force, the weak nuclear force, and electromagnetism into a single force. In these theories, the forces were thought to be unified at extremely high energies, similar to the conditions present during the early moments of the universe. Several GUTs were proposed, each with its predictions and consequences for particle physics and cosmology. One of the first GUTs was proposed by Howard Georgi and Sheldon Glashow in 1974. Their theory, known as the Georgi-Glashow model, sought to unify the electroweak force with the strong nuclear force. While the Georgi-Glashow model made successful predictions about particle interactions, it faced challenges in explaining certain phenomena and did not provide a complete framework for all four forces. Another influential GUT, the Pati-Salam model, was developed by Jogesh Pati and Abdus Salam in the 1970s. This model extended the Georgi-Glashow model to include an additional force, which was a combination of the strong nuclear force and the electroweak force. The Pati-Salam model offered a more comprehensive unification but still fell short of incorporating gravity into the unified framework. In the pursuit of a grand unified theory, physicists also explored the concept of supersymmetry, a theoretical framework that posits the existence of new, yet-undiscovered particles called superpartners. Supersymmetry suggested that every known particle in *the Standard Model of particle physics has a supersymmetric partner with different spin properties. The incorporation of supersymmetry into GUTs had the potential to provide a more complete and unified description of the fundamental forces. Supersymmetry also offered a natural candidate for dark matter, a mysterious form of matter that does not interact with light or other forms of electromagnetic radiation. While GUTs and supersymmetry showed promise, experimental evidence for these theories remained elusive, and some predicted particles, such as magnetic monopoles, had not been observed. The search for experimental confirmation of GUTs and supersymmetry has been ongoing, with particle accelerators and detectors designed to explore the high-energy regimes where the effects of unification might become visible. As of my last knowledge update in January 2022, no direct experimental evidence for GUTs or supersymmetry had been found. Despite the challenges and the lack of experimental confirmation, the pursuit of a unified theory of all four fundamental forces continues to be a driving force in theoretical physics. Modern theoretical frameworks, such as string theory and M-theory, aim to unify all fundamental forces, including gravity, within a single, coherent framework. These theories propose that the fundamental building blocks of the universe are not point-like particles but tiny, vibrating strings. String theory introduces extra dimensions beyond the familiar three spatial dimensions and one time dimension, which can potentially reconcile general relativity and quantum mechanics. While string theory has not yet been* experimentally verified and remains a topic of active research and debate, it represents one of the most ambitious attempts at unification in the history of physics. In summary, the quest for a unified theory of all four fundamental forces has been a profound and enduring endeavor in physics. Early attempts, such as the unification of electromagnetism and the development of grand unified theories, have expanded our understanding of the fundamental forces. Supersymmetry and string theory offer modern frameworks for potential unification, but experimental evidence remains a critical challenge. The pursuit of a unified theory continues to captivate the imagination of physicists and holds the promise of revealing the deepest secrets of the universe. The emergence of string theory represents a significant development in the quest for a unified theory of fundamental forces in physics. String theory is a theoretical framework that posits the fundamental building blocks of the universe as tiny, vibrating strings rather than point-like particles. It offers the tantalizing possibility of unifying all four known fundamental forces—gravity, electromagnetism, the strong nuclear force, and the weak nuclear force—within a single, coherent framework. The origins of string theory can be traced back to the late 1960s when theoretical physicists were grappling with the complexities of strong nuclear interactions and the behavior of hadrons, which are composite particles made up of quarks. The conventional approach to understanding these phenomena involved the development of quantum *chromodynamics (QCD), a theory that described the strong force in terms of quarks and gluons. While QCD was highly successful in describing the strong nuclear force, it left open the question of how to unify gravity with the other three fundamental forces. At the same time, a theory known as the S-matrix theory was being explored as a way to describe the scattering of particles. The S-matrix theory sought to find patterns and symmetries in the scattering amplitudes of particles, but it faced difficulties in accommodating the principles of quantum mechanics and special relativity. Amid these challenges, a breakthrough occurred when Gabriele Veneziano, an Italian physicist, formulated a mathematical formula known as the Veneziano amplitude in 1968. The Veneziano amplitude described certain properties of the strong nuclear force and had a mathematical form that exhibited a surprising symmetry. This mathematical discovery sparked interest among physicists and led to further investigations into similar mathematical structures. Leonard Susskind, Holger Bech Nielsen, and Joël Scherk were among the physicists who independently recognized the potential significance of the Veneziano amplitude and explored its implications. They noticed that the mathematical properties of the Veneziano amplitude were reminiscent of properties of vibrating strings. In 1970, Scherk and Susskind proposed a model of strings as fundamental objects to describe the strong nuclear force, leading to the birth of string theory. In this early version of string theory, the vibrating strings were open-ended and represented the building blocks of hadrons, such as mesons and baryons. However, this initial string theory was still primarily focused on the strong nuclear force and had not yet addressed the unification of all fundamental forces. The development of string theory received a significant boost with the discovery of certain mathematical structures known as dualities. These dualities revealed unexpected relationships between different versions of string theory and hinted at a more profound underlying structure. One of the key breakthroughs came with the realization that different string theories were related by various dualities, including T-duality and S-duality, which related string theories with different properties. These dualities suggested that string theory might represent a unified framework capable of describing all fundamental forces. In the mid-1980s, Edward Witten, a prominent physicist, played a pivotal role in advancing the understanding of string theory. Witten's insights and contributions helped unify various string theories into a single framework known as superstring theory. Superstring theory introduced the concept of supersymmetry, a theoretical symmetry between particles with different spin properties. Supersymmetry predicted the existence of new, yet-undiscovered particles called superpartners, which could potentially resolve some of the outstanding questions in particle physics. The development of superstring theory marked a significant step toward the unification of all fundamental forces. One of the groundbreaking features of string theory is its ability to consistently incorporate gravity into the framework. In classical physics, reconciling general relativity, which* describes gravity as the curvature of spacetime, with quantum mechanics, which deals with the behavior of particles at the smallest scales, has been a longstanding challenge. String theory offered a potential solution by describing gravity as the result of the vibrational modes of strings moving in a higher-dimensional spacetime. This approach hinted at the possibility of reconciling gravity with the other forces in a manner consistent with the principles of quantum mechanics. String theory also introduced the concept of extra dimensions beyond the familiar three spatial dimensions and one time dimension. These extra dimensions were proposed as the arena where strings move and interact, hidden from our everyday perception. The idea of extra dimensions raised intriguing possibilities for understanding the hierarchy of fundamental forces and the nature of the universe itself. As string theory continued to evolve, it led to the formulation of various versions, including Type I, Type IIA, Type IIB, heterotic, and M-theory. Each of these versions offered unique insights and perspectives on the underlying framework of string theory. One of the most remarkable developments was the proposal of M-theory by Edward Witten in the mid-1990s. M-theory, which stands for "magic," "mystery," or "membrane," depending on the interpretation, was conceived as a unifying framework that encompassed and connected the various string theories. M-theory introduced the concept of branes, extended objects in higher-dimensional spacetime, which played a central role in the theory. Branes allowed for a more comprehensive understanding of the dynamics of strings and their *interactions. The concept of duality, which had been instrumental in the development of string theory, took on new significance within the context of M-theory. M-theory introduced a web of dualities connecting different string theories and hinted at the existence of an underlying theory yet to be fully understood. Despite its elegance and potential, string theory faced several challenges. One of the primary challenges was the lack of experimental confirmation or direct empirical evidence for the existence of strings or the extra dimensions predicted by the theory. String theory also predicted the existence of supersymmetric particles, which had not been observed at the energies accessible to particle accelerators at the time. This absence of experimental verification posed a significant hurdle for the acceptance of string theory as a fundamental theory of nature. Additionally, the landscape of possible solutions within string theory, including the many possible compactifications of extra dimensions, made it difficult to make specific, testable predictions. Despite these challenges, string theory has continued to captivate the imagination of physicists and mathematicians. It has inspired new mathematical developments and fostered interdisciplinary collaborations between physicists and mathematicians. String theory's potential to unify all fundamental forces, reconcile gravity with quantum mechanics, and provide a comprehensive framework for understanding the universe remains a compelling motivation for ongoing research. The exploration of string theory has also led to insights into other areas of physics, such as black hole physics, quantum field theory, and holography. Moreover, the quest for experimental evidence and the search for supersymmetric particles continue at particle accelerators and in cosmological observations. In summary, the emergence of string theory represents a remarkable chapter in the history of theoretical physics. It has reshaped our understanding of the fundamental forces of nature, challenged conventional notions of particles, and provided a theoretical framework with profound implications for our perception of the universe. While string theory has yet to be experimentally confirmed and faces significant challenges, it remains one of the most ambitious and intriguing pursuits in the quest for a unified theory of everything.*

Chapter 3: String Theory Essentials

In the quest to understand the fundamental nature of the universe, theoretical physics has witnessed the emergence of a groundbreaking framework known as string theory. String theory represents a profound departure from conventional notions of particles and forces, offering a potential path toward a unified theory of all fundamental forces in the cosmos. At its heart, string theory proposes that the fundamental building blocks of the universe are not point-like particles but tiny, vibrating strings. These strings are minuscule and exist at scales far beyond our current experimental reach, making them a challenge to directly observe. The vibrational patterns of these strings are the source of all particles and interactions in the universe. The remarkable feature of string theory is its capacity to encompass all known fundamental forces, including gravity, electromagnetism, and the strong and weak nuclear forces, within a single theoretical framework. This vision of unification has been a long-standing aspiration in the field of physics, as it seeks to reconcile the seemingly disparate descriptions of these forces. To appreciate the significance of string theory, it is essential to grasp its historical context and the challenges it addresses. Throughout the 20th century, physicists made extraordinary strides in understanding the fundamental forces that govern the universe. Albert Einstein's theory of general relativity revolutionized our understanding of gravity by describing it as the curvature of spacetime. Meanwhile, quantum mechanics provided a comprehensive framework for understanding the behavior of particles and forces at the atomic and subatomic scales. Quantum field theory, a blend of quantum mechanics and special relativity, successfully described electromagnetism and the strong and weak nuclear forces. However, a fundamental issue remained unresolved: the incompatibility between general relativity and quantum mechanics. While both theories were remarkably successful in their respective domains, they presented fundamentally different descriptions of the physical world. General relativity described gravity as the curvature of spacetime, a continuous and geometric framework, whereas quantum mechanics described particles as discrete entities with well-defined properties. Efforts to reconcile these two theories, a quest for a theory of quantum gravity, were met with significant challenges. The fundamental nature of space and time, the behavior of particles at the smallest scales, and the structure of the universe itself all posed perplexing questions. One of the primary challenges was the behavior of gravity at the Planck scale, a tiny dimensionless number that characterizes the scale at which quantum gravitational effects become significant. At the Planck scale, the effects of quantum gravity are expected to manifest, but our current understanding is inadequate to describe this regime accurately. String theory emerged as a response to these challenges, offering a framework that inherently incorporates both quantum mechanics and gravity. The core idea of string theory is that the fundamental constituents of the universe are not point particles but tiny, vibrating strings. These strings come in various vibrational modes, each of which corresponds to a different particle or force. For instance, the lowest vibrational mode of a closed string corresponds to the graviton, a hypothetical particle associated with gravity. Higher vibrational modes yield particles such as photons (particles of light) or gluons (mediators of the strong nuclear force). This concept, known as the "stringy spectrum," provides a natural and unified description of particles and forces. String theory introduced the possibility of extra spatial dimensions beyond the familiar three dimensions of space. In string theory, these extra dimensions can be compactified or curled up, making them undetectable at macroscopic scales but influencing the behavior of strings at the quantum level. The idea of extra dimensions opened new avenues for understanding the hierarchy of fundamental forces and the structure of the universe. Moreover, string theory naturally incorporated the principles of supersymmetry, a theoretical symmetry that posits a relationship between particles with different spin properties. Supersymmetry predicts the existence of superpartners for every known particle, potentially resolving certain issues in particle physics and offering a candidate for dark matter. The development of string theory also brought forth the concept of dualities, which are profound symmetries connecting different versions of the theory. These dualities revealed that seemingly distinct string theories were, in fact, different manifestations of a more comprehensive framework. T-duality, for example, relates string theories with different spacetime geometries, while S-duality connects theories with different coupling constants. These dualities suggested that string theory represented a deeper and more interconnected structure than initially perceived. The culmination of these insights led to the formulation of a single overarching framework known as superstring theory, which unified various string theories into a coherent description of the universe. Superstring theory introduced the concept of branes, extended objects in higher-dimensional spacetime. These branes played a fundamental role in string theory and provided a new perspective on the dynamics of strings and their interactions. In particular, D-branes, a type of brane associated with open strings, became crucial in understanding the behavior of strings near branes. The rich mathematical structures within string theory also opened up new avenues for exploration in mathematics itself. Consequently, string theory fostered a deep and fruitful collaboration between physicists and mathematicians. Despite its elegance and potential, string theory faced significant challenges. One of the primary obstacles was the lack of experimental confirmation or direct empirical evidence for the existence of strings or the extra dimensions posited by the theory. The energy scales at which strings are thought to vibrate are orders of magnitude higher than those accessible by current particle accelerators. This made it difficult to directly observe the predicted stringy effects. Additionally, the landscape of possible solutions within string theory, including the numerous ways in which extra dimensions can be compactified, made it challenging to make specific, testable predictions. String theory also predicted the existence of supersymmetric particles, which had not been observed at the energies attainable by particle accelerators at the time. The absence of experimental verification posed a significant hurdle for the acceptance of string theory as a fundamental theory of nature. Despite these challenges, string theory continued to captivate the imagination of physicists and mathematicians alike. It represented a bold and audacious attempt to address some of the deepest questions in physics, such as the nature of spacetime, the origin of fundamental forces, and the structure of the universe. String theory's potential to unify all fundamental forces, reconcile gravity with quantum mechanics, and provide a comprehensive framework for understanding the cosmos remained a compelling motivation for ongoing research. Moreover, the exploration of string theory led to insights into other areas of physics, such as black hole physics, quantum field theory, and holography. The quest for experimental evidence and the search for supersymmetric particles continued at particle accelerators and in cosmological observations. In summary, the introduction to string theory represents a transformative chapter in the history of theoretical physics. It has reshaped our understanding of the fundamental forces of nature, challenged conventional notions of particles and forces, and provided a theoretical framework with profound implications for our perception of the universe. While string theory has yet to be experimentally confirmed and faces significant challenges, it remains one of the most ambitious and intriguing pursuits in the quest for a unified theory of everything. To delve deeper into the intricacies of string theory, it is essential to explore the concept of dimensions and vibrations that form the foundation of this groundbreaking framework. String theory introduces the notion that the fundamental building blocks of the universe are not point-like particles but tiny, vibrating strings. These strings are the essence of all particles and interactions, and their vibrational patterns give rise to the diverse phenomena observed in the cosmos. Central to string theory is the idea of extra dimensions, dimensions beyond the familiar three spatial dimensions and one time dimension. While our everyday experience is confined to these four dimensions, string theory posits the existence of additional, hidden dimensions. These extra dimensions are compactified or curled up to scales much smaller than can be directly observed, rendering them imperceptible in our macroscopic world. The presence of these compactified dimensions has profound implications for the behavior of strings and the unification of fundamental forces. In string theory, the total number of dimensions depends on the specific version of the theory under consideration. One common variant of string theory, known as Type IIB superstring theory, suggests the existence of ten dimensions, while others, like M-theory, propose eleven dimensions. These extra dimensions are a critical component of the mathematical and conceptual framework of string theory. To comprehend how strings vibrate and give rise to particles, it is essential to delve into the vibrational modes of these strings. Strings can vibrate in various ways, each corresponding to a distinct particle or force in the universe. The simplest vibrational mode corresponds to the graviton, a hypothetical particle associated with gravity. This mode represents the lowest energy state of a closed string and is crucial for understanding the behavior of gravity within the framework of string theory. Higher vibrational modes of strings yield different particles. For example, the first vibrational mode of a closed string corresponds to the dilaton, a particle that influences the strength of the gravitational force. The photon, which is the particle of light and electromagnetism, arises from a particular vibrational mode of open strings. Similarly, gluons, the mediators of the strong nuclear force, and W and Z bosons, which govern the weak nuclear force, are also generated by specific vibrational modes of strings. The vibrational patterns of strings are characterized by their frequency, amplitude, and polarization. These parameters determine the properties of the particles they give rise to, such as mass, charge, and spin. In essence, the diversity of particles and forces in the universe emerges from the symphony of vibrations played by these tiny strings. The concept of extra dimensions becomes particularly significant when considering how these dimensions influence string vibrations. In string theory, the geometry of spacetime, including the shape and size of extra dimensions, plays a crucial role in determining the vibrational spectrum of *strings. The compactified dimensions introduce quantization conditions on the vibrational modes, akin to the quantization of energy levels in a quantum mechanical system. The allowed vibrational frequencies are constrained by the geometry of these compact dimensions, giving rise to specific patterns of vibrational modes. This dependence on spacetime geometry underscores the intimate relationship between string theory and the geometry of extra dimensions. The vibrational modes of strings can be visualized as harmonics on a musical instrument. Just as plucking a stringed instrument at different positions generates distinct musical notes, the vibrational patterns of strings at various frequencies create different particles and forces. The harmonics produced by strings are not limited to single notes but encompass an entire orchestra of particles and interactions. Moreover, the concept of duality, a central theme in string theory, reveals that seemingly distinct string theories are intimately connected through transformations known as dualities. T-duality, for instance, relates string theories with different spacetime geometries by exchanging large and small dimensions. S-duality, on the other hand, connects theories with different coupling constants, revealing deeper symmetries in the fabric of string theory. These dualities hint at a profound unity underlying the diverse manifestations of string theory. The interplay between dimensions, vibrations, and dualities in string theory has sparked a profound transformation in our understanding of the fundamental forces of the universe. While the notion of extra dimensions may initially seem abstract or inaccessible, it is essential to recognize that they play a foundational role in string theory's attempt to unify all fundamental forces. The compactified dimensions offer a unique perspective on the hierarchy of forces and the structure of the universe itself. Moreover, the vibrational modes of strings provide a rich and unified description of particles and interactions, challenging conventional notions of point particles and forces. The symphony of vibrations orchestrated by strings offers a compelling narrative for the diversity of particles observed in the cosmos. Furthermore, the concept of duality unveils hidden connections between seemingly disparate aspects of string theory, suggesting a deeper and more interconnected framework. In summary, dimensions and vibrations are at the core of string theory's revolutionary approach to understanding the fundamental forces of the universe. These concepts introduce a profound shift in our perspective on particles and forces, emphasizing the role of tiny, vibrating strings in shaping the fabric of reality. The exploration of extra dimensions, the symphony of string vibrations, and the symmetries revealed by dualities continue to captivate the imagination of physicists and hold the promise of unveiling the deepest secrets of the cosmos.*

Chapter 4: Quantum Field Theory and Beyond

To embark on a journey through the realms of quantum field theory, it is imperative to establish a solid foundation in its fundamental principles and concepts. At its core, quantum field theory is a theoretical framework that extends the principles of quantum mechanics to fields, allowing us to describe and understand the behavior of particles and forces at the subatomic level. One of the pivotal ideas in quantum field theory is the concept of a field, which is a physical quantity that varies in space and time and has an associated quantum mechanical field operator. Fields can be thought of as entities that permeate all of space and time, and they play a fundamental role in our understanding of the behavior of particles. In quantum field theory, particles are not treated as isolated entities but are instead excitations or quanta of their respective fields. For example, the electromagnetic field gives rise to photons as its quanta, while the electron field produces electrons as its quanta. This perspective allows us to view particles as dynamic entities that interact with their corresponding fields and with one another. The mathematical formalism of quantum field theory is rooted in the principles of quantum mechanics and incorporates the principles of special relativity, making it a relativistic quantum theory. Special relativity, developed by Albert Einstein, introduced the idea that the laws of physics are the same for all observers *moving at constant velocities. In quantum field theory, this concept is essential for maintaining the consistency of physical laws in all reference frames and ensuring that physical predictions are consistent with experimental observations. One of the fundamental principles of quantum field theory is the quantization of fields. Quantization entails treating field values as operators that obey quantum mechanical commutation relations, leading to discrete energy levels or quanta. This quantization process results in the emergence of particles as excitations of these quantized fields. Quantum field theory employs a Lagrangian or Hamiltonian formalism to describe the dynamics of fields and particles. The Lagrangian is a mathematical function that summarizes the underlying physical interactions of the fields and particles, while the Hamiltonian represents the total energy of the system. These mathematical frameworks provide a systematic and rigorous way to derive the equations of motion governing the fields and particles. In quantum field theory, the behavior of fields and particles is described through a set of fundamental principles and equations. One of the central equations in quantum field theory is the Klein-Gordon equation, which governs the behavior of scalar fields. Scalar fields are those that do not carry any intrinsic angular momentum, also known as spin. The Klein-Gordon equation describes how these scalar fields evolve in spacetime, taking into account their mass and interactions. Another essential equation is the Dirac equation, which describes the behavior of fermionic fields. Fermions are particles with half-integer* spin, such as electrons and quarks. The Dirac equation provides a quantum description of fermions, incorporating their spin and interactions with other fields. For example, it describes how electrons move and interact with the electromagnetic field, giving rise to phenomena like electric conductivity. The quantization of fields in quantum field theory also leads to the concept of particle creation and annihilation. Particles can be created from the vacuum, which is the lowest energy state of a field, and they can annihilate, returning their energy to the field. This process is fundamental to understanding phenomena such as particle scattering and the creation and annihilation of particle-antiparticle pairs in particle accelerators. One of the significant achievements of quantum field theory is its ability to provide a unified description of particles and forces within a single theoretical framework. In this framework, particles are treated as excitations of their corresponding fields, and the interactions between particles are mediated by the exchange of other particles, known as force carriers. For example, the electromagnetic force is described by the exchange of photons between charged particles, while the strong nuclear force is mediated by gluons. The unification of particles and forces in quantum field theory has been a profound and successful endeavor, leading to the development of the Standard Model of particle physics. The Standard Model is a comprehensive theory that describes the fundamental particles of the universe and their interactions through the electromagnetic, weak nuclear, and strong nuclear forces. It has been validated through numerous experiments and is one of the most successful theories in the history of physics. However, the Standard Model is not without its limitations. It does not incorporate gravity, and it leaves many questions unanswered, such as the nature of dark matter and the unification of all fundamental forces. As a result, physicists continue to explore and extend the principles of quantum field theory in search of a more comprehensive theory that encompasses all known forces, including gravity. The development of quantum field theory has also had a profound impact on our understanding of the behavior of particles at high energies and in extreme conditions, such as those encountered in the early universe or in the vicinity of black holes. Quantum field theory provides essential tools for studying phenomena such as particle collisions at particle accelerators, the behavior of matter at high temperatures and densities, and the quantum properties of spacetime itself. The concept of renormalization, a technique to remove infinities that arise in certain calculations, has been crucial for making quantum field theory predictions consistent and finite. Moreover, the theory of quantum electrodynamics (QED), which describes the electromagnetic interactions of particles and fields, is one of the most precisely tested theories in physics, with predictions confirmed to incredible precision in experiments. In summary, quantum field theory serves as the theoretical framework that underpins our understanding of the behavior of particles and fields at the most fundamental level. It unifies the principles of quantum mechanics and special relativity, *describing particles as excitations of fields and interactions through the exchange of force carriers. Quantum field theory has yielded profound insights into the nature of matter and forces, leading to the development of the Standard Model and informing our understanding of the behavior of particles in a wide range of physical phenomena. As physicists continue their exploration of the quantum realm, quantum field theory remains an indispensable tool for unraveling the mysteries of the universe. To embark on a journey through the world of quantum field theory and its profound connection to particle physics, it is essential to grasp the intricate interplay between these two fields of study. Quantum field theory represents the theoretical framework that underlies our understanding of the behavior of particles and forces at the subatomic level. At its core, it extends the principles of quantum mechanics to fields, allowing us to describe and comprehend the behavior of particles in a quantum mechanical context. The development of quantum field theory was driven by the need to reconcile quantum mechanics with the principles of special relativity, as the latter introduced the concept that the laws of physics should be invariant for all observers moving at constant velocities. This unification was essential for creating a consistent description of particles and forces in the universe. In quantum field theory, particles are not treated as isolated entities but are considered excitations or quanta of their respective fields. Fields, in this context, are physical quantities that vary in space and time and have associated quantum field operators.*

These fields pervade all of spacetime, and their interactions with particles and with one another govern the dynamics of the subatomic world. An essential aspect of quantum field theory is the quantization of fields, which involves treating field values as operators that satisfy specific quantum mechanical commutation relations. This quantization process results in the emergence of particles as discrete energy levels or quanta of their respective fields. For instance, the electromagnetic field gives rise to photons as its quanta, while the electron field produces electrons as its quanta. This perspective challenges the classical notion of particles as distinct entities and emphasizes their dynamic nature as excitations of fields. Furthermore, quantum field theory incorporates the principles of special relativity into its mathematical framework, making it a relativistic quantum theory. The fusion of quantum mechanics and special relativity is crucial for ensuring that the laws of physics are consistent in all inertial reference frames and for describing particles moving at significant fractions of the speed of light. To describe the dynamics of fields and particles within quantum field theory, mathematical formalisms based on either Lagrangian or Hamiltonian formulations are employed. The Lagrangian is a mathematical function that encapsulates the underlying physical interactions of fields and particles, while the Hamiltonian represents the total energy of the system. These formalisms provide systematic and rigorous means of deriving the equations of motion governing the fields and particles. A fundamental principle in quantum field theory is the concept of particle creation and annihilation. Particles can be created from the vacuum, which represents the lowest energy state of a field, and they can also annihilate, returning their energy to the field. This process plays a fundamental role in understanding various phenomena, such as particle scattering and the creation and annihilation of particle-antiparticle pairs in particle accelerators. The mathematical equations that govern quantum field theory depend on the nature of the fields being considered. One of the central equations in quantum field theory is the Klein-Gordon equation, which describes the behavior of scalar fields. Scalar fields are those without intrinsic angular momentum, also known as spin. The Klein-Gordon equation takes into account the mass and interactions of scalar fields and elucidates their dynamics in spacetime. Another pivotal equation is the Dirac equation, which pertains to fermionic fields. Fermions are particles characterized by half-integer spin, such as electrons and quarks. The Dirac equation offers a quantum description of fermions, accounting for their spin and interactions with other fields. It provides insights into phenomena like electric conductivity, where electrons interact with the electromagnetic field. The quantization of fields in quantum field theory also gives rise to the concept of force carriers. Particles interact with one another through the exchange of other particles, known as force carriers. For instance, the electromagnetic force is mediated by the exchange of photons between charged particles. Similarly, the strong nuclear force is governed by the exchange of gluons, and the weak nuclear force *involves the exchange of W and Z bosons. This unification of particles and forces within quantum field theory has been a remarkable achievement, culminating in the development of the Standard Model of particle physics. The Standard Model is a comprehensive theory that describes the fundamental particles of the universe and their interactions through the electromagnetic, weak nuclear, and strong nuclear forces. It has been validated through extensive experiments and is one of the most successful theories in the history of physics. Nonetheless, the Standard Model has its limitations. It does not incorporate gravity, and it leaves unresolved questions such as the nature of dark matter and the unification of all fundamental forces. As a result, physicists continue to explore and extend the principles of quantum field theory in search of a more comprehensive theory that encompasses all known forces, including gravity. Quantum field theory has not only revolutionized our understanding of particles and forces but has also had a profound impact on various areas of physics. It provides essential tools for studying high-energy phenomena and extreme conditions, such as those found in the early universe or near black holes. The theory of quantum electrodynamics (QED), which describes the electromagnetic interactions of particles and fields, is one of the most precisely tested theories in physics, with predictions confirmed to remarkable precision in experiments. Quantum field theory is instrumental in elucidating phenomena like particle collisions at particle accelerators, the behavior of matter at high temperatures and densities, and the quantum properties of spacetime itself. The concept of renormalization, a technique to remove infinities that arise in certain calculations, has been fundamental in making quantum field theory predictions finite and consistent. In summary, quantum field theory serves as the theoretical framework that forms the bedrock of our understanding of the subatomic world. It unites the principles of quantum mechanics and special relativity, portraying particles as manifestations of fields and interactions through the exchange of force carriers. Quantum field theory has provided profound insights into the nature of matter and forces, culminating in the development of the Standard Model and guiding our comprehension of particles in a wide array of physical phenomena. As physicists continue their exploration of the quantum realm, quantum field theory remains an indispensable instrument for unraveling the enigmas of the universe.*

Chapter 5: Quantum Computing Basics

Quantum computing represents a revolutionary paradigm in the world of computation, one that harnesses the profound and often counterintuitive principles of quantum mechanics to perform computations that were previously thought to be infeasible. While classical computers rely on bits to encode and process information as 0s and 1s, quantum computers utilize quantum bits or qubits that can exist in superpositions of 0 and 1 states, allowing them to perform vast numbers of calculations simultaneously. The idea of quantum computing dates back to the early 1980s when physicist Richard Feynman suggested that simulating quantum systems using classical computers was an immensely challenging task, leading to the concept of quantum computers. In 1985, David Deutsch formalized the concept by proposing the theoretical framework for a quantum Turing machine. However, it wasn't until the 1990s that the field gained substantial momentum, thanks to the groundbreaking work of Peter Shor and Lov Grover. Shor's algorithm demonstrated the potential of quantum computers to factor large numbers exponentially faster than classical computers, posing a significant threat to classical encryption schemes. Grover's algorithm, on the other hand, showcased quantum computing's ability to perform unstructured search problems quadratically faster than classical algorithms. Quantum computing leverages the fundamental principles of quantum mechanics, which govern the behavior of particles at the smallest scales. One of these principles is superposition, which allows quantum bits or qubits to exist in multiple states simultaneously. In classical computing, a bit can be either 0 or 1, but in quantum computing, a qubit can be in a superposition of both states, effectively performing multiple calculations in parallel. Another key quantum principle is entanglement, where the quantum states of two or more particles become correlated in such a way that the state of one particle instantly influences the state of the others, even if they are separated by vast distances. Entanglement is a fundamental resource in quantum computing, enabling the creation of quantum gates that perform operations on qubits. Quantum gates are the quantum analogs of classical logic gates and are essential for manipulating qubits to perform computations. The third critical principle is quantum interference, which allows quantum computers to exploit the probabilistic nature of quantum mechanics to enhance the likelihood of obtaining the correct solution in certain algorithms. These principles make quantum computing a fundamentally different approach to computation than classical computing, with the potential to solve problems that are currently beyond the reach of even the most powerful supercomputers. Quantum computers hold the promise of revolutionizing various fields, including cryptography, optimization, drug discovery, materials science, and artificial intelligence. In cryptography, for instance, quantum computers could break widely used encryption methods, prompting the need for quantum-safe encryption algorithms. On the other hand, they could also enable secure communication through the development of quantum cryptography protocols, such as quantum key distribution. In optimization problems, quantum computing could provide more efficient solutions for tasks like route optimization, portfolio optimization, and supply chain management. Quantum computers could also accelerate the discovery of new materials with unique properties, potentially revolutionizing industries like electronics and energy storage. Moreover, in the field of artificial intelligence, quantum computing could enhance machine learning algorithms by significantly speeding up the training and optimization of complex models. While the potential of quantum computing is vast, building practical and scalable quantum computers remains a formidable challenge. Quantum systems are highly sensitive to their environment, leading to a phenomenon known as decoherence, where qubits lose their quantum properties and behave classically. Efforts are underway to develop error-correcting codes and fault-tolerant quantum computing architectures to mitigate the effects of decoherence. Another challenge is the need to maintain qubits at extremely low temperatures, close to absolute zero, to reduce thermal noise and decoherence. Various physical platforms are being explored for building quantum computers, including superconducting qubits, trapped ions, and topological qubits. Each platform has its advantages and limitations, and researchers are actively working to overcome technical hurdles and improve qubit performance. The development of quantum software is also crucial for harnessing the power of quantum computers. Quantum programming languages, such as Qiskit, Cirq, and Quipper, have been developed to enable researchers and developers to write quantum algorithms and simulations. These languages provide abstractions and libraries that make it easier to work with qubits and quantum gates. Moreover, quantum cloud platforms, like IBM Quantum Experience and Google Quantum Computing, allow users to access quantum hardware over the internet, democratizing access to quantum computing resources. Quantum algorithms, such as Shor's algorithm for factoring large numbers and Grover's algorithm for unstructured search, continue to be refined and adapted for practical use cases. Quantum machine learning, quantum cryptography, and quantum chemistry simulations are among the emerging applications of quantum computing that hold significant promise. As the field of quantum computing advances, it is expected to have a profound impact on industries and scientific disciplines. Quantum computers have the potential to revolutionize cryptography, optimization, materials science, drug discovery, and artificial intelligence. However, many technical challenges must be overcome to build practical and scalable quantum computers, including mitigating decoherence and improving qubit performance. Quantum software development is also essential, with quantum programming languages and cloud platforms enabling researchers and developers to harness quantum computing resources. While quantum computing is still in its early stages, its potential is nothing short of transformative, and it promises to reshape the landscape of computational capabilities in the coming decades. The future of quantum computing is brimming with exciting possibilities, and researchers around the world are pushing the boundaries of our understanding of quantum mechanics to unlock the full potential of this revolutionary technology. To delve into the fascinating world of quantum computing, it is essential to grasp the fundamental building blocks that form the bedrock of this revolutionary field, namely quantum bits or qubits and quantum gates. Qubits represent the quantum analogs of classical bits, and they serve as the fundamental units of information in quantum computing. Unlike classical bits, which can only exist in one of two states, 0 or 1, qubits can exist in superpositions of these states, allowing them to represent multiple values simultaneously. This property is a cornerstone of quantum computing's power, as it enables qubits to perform parallel calculations that would be infeasible for classical bits. Furthermore, qubits possess another remarkable property known as entanglement, where the quantum states of qubits become correlated in such a way that the state of one qubit instantly influences the state of another, even if they are separated by vast distances. Entanglement is a crucial resource in quantum computing, enabling the creation of quantum gates that perform operations on qubits. Quantum gates are the quantum counterparts of classical logic gates, such as AND, OR, and NOT gates, and they are *fundamental for manipulating qubits and performing quantum computations. These gates are responsible for executing various quantum algorithms and solving complex problems with quantum computers. In quantum computing, qubits are represented as vectors in a complex vector space, and their states can be described using mathematical notation. A qubit can exist in a quantum state represented as $|\psi\rangle$, where the vertical bars denote a quantum state, and ψ represents the state vector. The state vector can be expressed as a linear combination of the basis states $|0\rangle$ and $|1\rangle$, where $|0\rangle$ corresponds to the state 0 and $|1\rangle$ corresponds to the state 1. The superposition property of qubits allows them to be in a linear combination of both $|0\rangle$ and $|1\rangle$ states simultaneously. Mathematically, a qubit in a superposition state can be represented as $|\psi\rangle = \alpha|0\rangle + \beta|1\rangle$, where α and β are complex numbers that determine the probability amplitudes of measuring the qubit in states $|0\rangle$ and $|1\rangle$. These probability amplitudes square to yield the probabilities of measuring the qubit in the respective states. The requirement that the sum of the probabilities of all possible measurement outcomes must equal 1 ensures the normalization of the qubit's state vector. The concept of qubits and their superposition forms the foundation for the quantum parallelism that underlies quantum computing's potential for speedup in specific tasks. In classical computing, performing a calculation on n bits requires a linear progression of operations, whereas in quantum computing, a quantum algorithm can operate on a superposition of 2^n states* simultaneously, offering exponential speedup for certain problems. Quantum gates are the tools that allow us to manipulate qubits and perform operations on them. Quantum gates are represented by unitary matrices, which are mathematical operators that preserve the normalization of the qubit's state vector. These gates transform the state of qubits according to specific rules and can be combined to create complex quantum circuits. One of the simplest quantum gates is the Pauli-X gate, often referred to as the quantum NOT gate. The Pauli-X gate acts as a bit-flip gate, transforming the state $|0\rangle$ into $|1\rangle$ and vice versa. Another essential gate is the Pauli-Y gate, which combines a bit-flip with a phase flip, and the Pauli-Z gate, which induces a phase flip. These gates, along with the identity gate (I), form the Pauli group, which plays a fundamental role in quantum computing. In addition to the Pauli gates, quantum computing employs various other gates, including the Hadamard gate (H) and the phase gate (S). The Hadamard gate is responsible for creating superposition states and is a key ingredient in many quantum algorithms. The phase gate introduces a phase shift to the qubit's state, altering the probability amplitudes. One of the most influential quantum gates is the controlled-NOT gate (CNOT), which is a two-qubit gate that acts on a target qubit based on the state of a control qubit. The CNOT gate enables entanglement between qubits and is a fundamental building block for constructing quantum circuits. Quantum circuits are sequences of quantum gates that manipulate qubits to perform specific computations or solve problems. These *circuits can be visualized as networks of interconnected gates, with qubits entering the circuit in specific initial states and exiting after undergoing various transformations. Quantum algorithms are designed to take advantage of the unique properties of qubits and quantum gates to solve problems more efficiently than classical algorithms. One of the most famous quantum algorithms is Shor's algorithm, which can factor large numbers exponentially faster than the best-known classical algorithms. This poses a significant threat to classical encryption schemes, as many cryptographic protocols rely on the difficulty of factoring large numbers. Another influential quantum algorithm is Grover's algorithm, which accelerates unstructured search problems quadratically faster than classical algorithms. Grover's algorithm has applications in database search and optimization. Quantum computing also has the potential to revolutionize fields such as cryptography, optimization, drug discovery, materials science, and artificial intelligence. For instance, quantum computers could break widely used encryption methods, necessitating the development of quantum-safe encryption algorithms. On the other hand, they could enable secure communication through quantum cryptography protocols like quantum key distribution. In optimization problems, quantum computing could provide more efficient solutions for tasks like route optimization, portfolio optimization, and supply chain management. Moreover, in the field of artificial intelligence, quantum computing could enhance machine learning algorithms by significantly speeding* up the training and optimization of complex models. While quantum computing holds immense promise, building practical and scalable quantum computers remains a significant challenge. Qubits are highly sensitive to their environment, leading to decoherence, where they lose their quantum properties and behave classically. Error-correcting codes and fault-tolerant quantum computing architectures are being developed to mitigate the effects of decoherence. Physical platforms for building quantum computers, such as superconducting qubits, trapped ions, and topological qubits, each have their advantages and limitations, and researchers are actively working to improve qubit performance. Quantum software development is also crucial, with quantum programming languages and cloud platforms enabling researchers and developers to harness quantum computing resources. Quantum algorithms continue to be refined and adapted for practical use cases in areas like quantum machine learning, quantum cryptography, and quantum chemistry simulations. In summary, quantum bits or qubits, along with quantum gates, are the fundamental elements that power quantum computing. Qubits can exist in superpositions of states, offering the potential for exponential speedup in certain tasks. Quantum gates manipulate qubits according to specific rules and create complex quantum circuits. Quantum algorithms leverage these properties to perform computations more efficiently than classical algorithms. While quantum computing has the potential to revolutionize various fields, building practical and scalable quantum *computers remains a formidable challenge, with ongoing efforts to overcome technical hurdles and develop quantum software.*

Chapter 6: Quantum Algorithms and Quantum Supremacy

To explore the transformative potential of quantum computing fully, it is crucial to delve into quantum algorithms and their myriad applications across various domains of science, industry, and technology. Quantum algorithms are specialized algorithms designed to harness the unique properties of quantum bits (qubits) and quantum gates to solve problems more efficiently than classical algorithms. One of the most celebrated quantum algorithms is Shor's algorithm, which has the remarkable ability to factor large numbers exponentially faster than the best-known classical algorithms. This property poses a significant challenge to classical encryption methods that rely on the difficulty of factoring large numbers for security. Shor's algorithm threatens the security of widely used encryption schemes, such as RSA, and has prompted the development of quantum-resistant encryption techniques. Another influential quantum algorithm is Grover's algorithm, which offers a quadratic speedup in solving unstructured search problems compared to classical algorithms. Grover's algorithm has diverse applications, including database search, optimization, and solving black-box problems. Quantum algorithms have also made significant strides in the field of quantum chemistry, where they promise to revolutionize the simulation of molecular and chemical systems.

Efficiently simulating the behavior of molecules and materials is a challenging computational task, as it involves solving complex quantum mechanical equations. Quantum algorithms like the quantum variational eigensolver (QVE) and the quantum phase estimation algorithm hold the potential to provide accurate solutions to quantum chemistry problems that are intractable for classical computers. These quantum algorithms are particularly valuable for drug discovery, material design, and understanding the properties of molecules. The field of quantum machine learning is another domain where quantum algorithms are poised to make a substantial impact. Quantum machine learning leverages quantum computing's parallelism and interference to accelerate tasks such as data classification, clustering, and regression. Quantum algorithms like the quantum support vector machine (QSVM) and quantum principal component analysis (PCA) aim to outperform their classical counterparts in machine learning tasks. Quantum-enhanced optimization algorithms are of significant interest in various industries, including finance, logistics, and supply chain management. Quantum algorithms like the quantum approximate optimization algorithm (QAOA) and the quantum annealing algorithm have the potential to find more efficient solutions to complex optimization problems. These algorithms can be applied to tasks like portfolio optimization, route planning, and resource allocation. Furthermore, quantum algorithms have the potential to transform the field of cryptography by offering new approaches to secure communication and data protection. Quantum cryptography protocols, such as quantum key distribution (QKD), use the fundamental principles of quantum mechanics to provide secure communication channels. These protocols leverage the no-cloning theorem and the uncertainty principle to ensure that any eavesdropping attempts would disrupt the quantum state of the transmitted information, alerting the parties involved. Quantum-resistant encryption techniques are also being developed to withstand attacks from quantum computers that could break classical encryption methods. Quantum algorithms have applications in the field of artificial intelligence (AI) as well. Quantum machine learning algorithms aim to enhance the training and optimization of complex AI models. Quantum algorithms can potentially speed up tasks like training deep neural networks, optimizing large-scale machine learning models, and solving combinatorial optimization problems often encountered in AI. Furthermore, quantum algorithms have garnered attention in the realm of quantum simulations, where they promise to provide insights into complex physical systems that are challenging to study using classical methods. Quantum simulations enable the exploration of quantum materials, condensed matter physics, and the behavior of particles in extreme conditions. Quantum algorithms, such as the quantum phase estimation algorithm and the quantum matrix inversion algorithm, can simulate quantum systems efficiently. These simulations are invaluable for advancing our understanding of fundamental physics and driving *innovations in materials science. Quantum computing also holds the potential to revolutionize the field of finance and risk assessment. Quantum algorithms for pricing financial derivatives, optimizing investment portfolios, and simulating complex financial models could provide more accurate and efficient solutions. Moreover, quantum algorithms have applications in supply chain optimization, where they can help optimize logistics, inventory management, and resource allocation in a highly interconnected world. The field of cryptography is poised for transformation as well, with quantum-resistant encryption techniques and quantum cryptography protocols offering enhanced security. Quantum computing's potential to disrupt existing cryptographic methods necessitates the development of new cryptographic protocols that can withstand attacks from quantum computers. Quantum cryptography, including quantum key distribution (QKD) and quantum secure communication, leverages the principles of quantum mechanics to provide secure channels for communication. These protocols rely on the fundamental properties of quantum states to detect any unauthorized attempts to intercept or tamper with transmitted information. Quantum-resistant encryption methods, such as lattice-based cryptography and hash-based cryptography, are being researched and developed to replace classical encryption methods vulnerable to quantum attacks. The advent of quantum computing has opened up a new frontier in the realm of optimization, offering the promise of more efficient solutions to complex problems. Quantum optimization algorithms, like the quantum approximate optimization algorithm (QAOA) and quantum annealing algorithms, have applications in various domains. They can optimize routes for delivery trucks, find the most cost-effective manufacturing processes, and allocate resources efficiently. Quantum algorithms are poised to drive advancements in AI and machine learning by speeding up training processes, enhancing optimization tasks, and solving combinatorial optimization problems that underlie many AI applications. Quantum machine learning algorithms, such as the quantum support vector machine (QSVM) and quantum principal component analysis (PCA), are at the forefront of these developments. Quantum simulations represent a crucial application of quantum algorithms, allowing researchers to gain insights into complex physical systems and simulate the behavior of particles under extreme conditions. Quantum algorithms for simulations can advance our understanding of materials science, condensed matter physics, and quantum materials. The financial industry stands to benefit from quantum computing, with quantum algorithms for pricing financial derivatives, optimizing investment portfolios, and simulating financial models offering potential advantages. Supply chain optimization is another area where quantum computing can play a transformative role, optimizing logistics, inventory management, and resource allocation in a highly interconnected world. In summary, quantum algorithms are at the heart of quantum computing's promise to revolutionize various fields, including cryptography, optimization, finance, AI, materials science, and supply chain management. These algorithms leverage the unique properties of quantum bits (qubits) and quantum gates to solve complex problems more efficiently than classical algorithms. While quantum computing is still in its early stages, ongoing research and development efforts hold the potential to unlock the full power of quantum algorithms and their diverse applications in our increasingly interconnected and data-driven world. To understand the concept of quantum supremacy, we must embark on a journey into the realm of quantum computing's potential and the milestones that mark its progress. Quantum supremacy refers to the point at which a quantum computer can perform a specific task or solve a problem faster than the most advanced classical supercomputers available. This milestone represents a watershed moment in the development of quantum computing, showcasing its ability to outperform classical computing in a real-world application. Quantum supremacy is not merely a theoretical concept; it has become a tangible goal for researchers and organizations in the field of quantum computing. The path to achieving quantum supremacy involves overcoming formidable challenges, including the development of reliable and error-corrected quantum hardware, the creation of specialized quantum algorithms, and the demonstration of quantum advantage in practical computations. Quantum hardware forms the foundation of quantum computing, and it is composed of qubits, quantum gates, and the infrastructure required to maintain and manipulate quantum states. Building qubits that are stable, error-resistant, and scalable is a complex engineering task. Various physical platforms are being explored for qubits, including superconducting qubits, trapped ions, and topological qubits, each with its strengths and challenges. Error-correcting codes and fault-tolerant quantum computing architectures are being developed to mitigate the effects of decoherence and errors in quantum hardware. These error-correction techniques are essential for achieving reliable and scalable quantum computing systems. Quantum algorithms play a crucial role in realizing quantum supremacy by harnessing the unique properties of quantum bits and quantum gates to solve problems efficiently. Shor's algorithm and Grover's algorithm are among the most famous quantum algorithms that have demonstrated quantum advantage in specific domains. Shor's algorithm can factor large numbers exponentially faster than classical algorithms, posing a threat to classical encryption schemes. Grover's algorithm accelerates unstructured search problems quadratically faster than classical algorithms, offering benefits in tasks like database search and optimization. Quantum supremacy, however, extends beyond these algorithms and encompasses a broader range of applications where quantum computers can excel. Demonstrating quantum supremacy requires executing a computation that is practically infeasible for classical supercomputers to replicate within a reasonable time frame. Google's achievement of quantum supremacy in 2019 marked a significant milestone in the field. They used their 53-*

qubit quantum processor, Sycamore, to perform a task known as random circuit sampling, demonstrating that their quantum computer could complete the task much faster than any classical computer. IBM challenged Google's claim, arguing that classical supercomputers could simulate the quantum task's output more efficiently. This debate sparked discussions in the scientific community about the definition and demonstration of quantum supremacy. Regardless of this debate, achieving quantum supremacy is a remarkable achievement that showcases the progress made in quantum computing. Quantum computers are not yet ready to replace classical computers for everyday tasks, as they still face significant engineering challenges and limitations. However, they hold immense potential for solving specific problems faster and more efficiently, making them a valuable tool in fields like cryptography, optimization, materials science, and artificial intelligence. Quantum computing is still in its infancy, and researchers continue to explore ways to improve qubit performance, error correction, and quantum algorithms. In the quest for quantum supremacy, researchers are not only focusing on hardware but also on software development. Quantum programming languages, such as Qiskit and Cirq, provide tools and libraries for writing quantum algorithms and simulations. Quantum cloud platforms, like IBM Quantum Experience and Google Quantum Computing, enable users to access quantum hardware and test their algorithms remotely. Quantum supremacy is not the endpoint of quantum computing but rather a milestone on the journey to harness its full potential. Researchers are now striving to demonstrate quantum advantage in more practical and impactful applications. Quantum machine learning, quantum cryptography, and quantum simulations are some of the areas where quantum computers can make a significant difference. Quantum machine learning algorithms aim to enhance the training and optimization of machine learning models, offering speedup in various AI tasks. Quantum cryptography protocols, such as quantum key distribution (QKD), provide secure communication channels immune to quantum attacks. Quantum simulations enable the study of complex physical systems, materials, and quantum chemistry with unprecedented accuracy. The pursuit of quantum supremacy has ignited competition among leading technology companies and research institutions. They invest in quantum research, hardware development, and algorithm optimization to push the boundaries of quantum computing. While achieving quantum supremacy is a significant milestone, the ultimate goal is to develop practical and scalable quantum computers that can address real-world challenges. Quantum computing's potential to disrupt various industries, including finance, healthcare, and logistics, is driving innovation and collaboration across sectors. Quantum computers are not just faster versions of classical computers; they operate on fundamentally different principles, leveraging the properties of quantum bits to perform computations that were previously impossible or infeasible. As we continue our exploration of *quantum computing and the quest for quantum supremacy, it is essential to appreciate the multidisciplinary efforts of scientists, engineers, and researchers who are shaping the future of this transformative technology. Quantum supremacy represents a significant step forward, but the journey is far from over, and the potential applications of quantum computing are vast and promising. The path to quantum supremacy is not linear; it involves overcoming challenges, revising definitions, and continually pushing the boundaries of what is possible in the world of quantum computing. In the chapters ahead, we will delve deeper into the principles, algorithms, and applications of quantum computing, exploring how this revolutionary technology is poised to reshape our world.*

Chapter 7: Quantum Information and Entanglement

To grasp the intricacies of quantum information theory, we must embark on a journey through the quantum realm, where the laws of classical information theory no longer apply, and the principles of quantum mechanics reign supreme. Information theory, a field initially formulated by Claude Shannon in the mid-20th century, established a framework for understanding the transmission and processing of classical information. In classical information theory, information is represented as bits, which can exist in one of two states: 0 or 1. These bits serve as the fundamental units of information, forming the basis for classical communication and computation. However, when we transition to the quantum realm, we encounter a profound shift in our understanding of information. Quantum information theory arises from the marriage of quantum mechanics and information theory, introducing a new perspective on how information can be stored, transmitted, and manipulated in the quantum world. At the heart of quantum information theory are quantum bits, or qubits, which replace classical bits as the fundamental units of information. Unlike classical bits, qubits can exist in superpositions of states, allowing them to represent multiple values simultaneously. This property is a cornerstone of quantum information processing, as it enables qubits to perform parallel computations that classical bits cannot replicate.

Furthermore, qubits possess another remarkable property known as entanglement, where the quantum states of qubits become correlated in such a way that the state of one qubit instantly influences the state of another, even if they are separated by vast distances. Entanglement is a crucial resource in quantum information theory, enabling the creation of quantum communication protocols and quantum algorithms that exploit this interconnectedness to achieve novel feats. One of the earliest and most celebrated concepts in quantum information theory is quantum teleportation, a phenomenon that challenges our classical intuitions about transmitting information. In quantum teleportation, the state of one qubit can be transmitted from one location to another without physically moving the qubit itself. This process relies on entanglement and quantum entanglement swapping, allowing the instantaneous transfer of quantum information over long distances. Quantum teleportation has practical implications for secure quantum communication, such as quantum key distribution, where the transfer of quantum states guarantees the security of the communication channel. Another foundational concept in quantum information theory is quantum superposition, which enables qubits to exist in a linear combination of basis states, each with a probability amplitude. This property allows quantum algorithms to perform computations in parallel by exploring multiple paths of computation simultaneously. Quantum superposition is harnessed in quantum algorithms like Grover's algorithm, which accelerates unstructured search problems, and quantum Fourier transform, a key component of Shor's algorithm for integer factorization. These quantum algorithms showcase the power of quantum superposition in solving problems faster than classical counterparts. Quantum information theory also introduces the concept of quantum gates, which are the quantum counterparts of classical logic gates. Quantum gates manipulate qubits to perform specific operations, and they play a central role in quantum algorithms and quantum circuits. Common quantum gates include the Pauli-X gate, which acts as a quantum NOT gate, and the Hadamard gate, which creates superposition states. The controlled-NOT gate (CNOT) enables entanglement between qubits and is fundamental for creating quantum circuits. Quantum gates are essential tools for designing quantum algorithms that harness the unique properties of qubits. Quantum entanglement, another hallmark of quantum information theory, is a phenomenon where the quantum states of two or more qubits become correlated in such a way that measuring one qubit instantly determines the state of the others, regardless of the distance between them. This non-local correlation challenges classical notions of separability and has been famously described as "spooky action at a distance" by Albert Einstein. Entanglement is a valuable resource in quantum communication and quantum computing, enabling secure quantum key distribution and facilitating the creation of quantum circuits that exploit quantum parallelism. In quantum cryptography, researchers have developed quantum key distribution (QKD) protocols that leverage the principles of quantum mechanics to provide secure communication channels. QKD protocols, such as the BBM92 protocol and the E91 protocol, use entangled qubits to enable secure key exchange between two parties. This secure key can then be used for encrypted communication, ensuring the confidentiality of transmitted information. Quantum cryptography promises unbreakable encryption, as any eavesdropping attempts would disrupt the quantum state of the transmitted qubits, alerting the parties involved. Quantum information theory also extends to the realm of quantum computation, where quantum algorithms leverage the unique properties of qubits to solve problems more efficiently than classical algorithms. Shor's algorithm, for instance, can factor large numbers exponentially faster than the best-known classical algorithms, posing a significant threat to classical encryption methods. Grover's algorithm accelerates unstructured search problems quadratically faster than classical algorithms, making it valuable for tasks like database search and optimization. Quantum computing has the potential to revolutionize various fields, including materials science, drug discovery, optimization, and artificial intelligence. Simulating the behavior of molecules and materials is a computationally intensive task, but quantum algorithms like the quantum variational eigensolver (QVE) and the quantum phase estimation algorithm promise to provide accurate solutions for quantum chemistry problems that are intractable for classical computers. Quantum machine learning algorithms aim to enhance the training and optimization of machine learning models by *leveraging quantum parallelism and interference. Quantum cryptography protocols, such as quantum key distribution (QKD), provide secure communication channels immune to quantum attacks. Quantum simulations enable the study of complex physical systems, materials, and quantum chemistry with unprecedented accuracy. Quantum algorithms for optimization have applications in finance, logistics, supply chain management, and AI, offering more efficient solutions to complex problems. Quantum information theory is at the forefront of quantum computing's potential to revolutionize industries and solve problems that were once considered insurmountable. It bridges the gap between the principles of quantum mechanics and the processing of information, opening new avenues for secure communication, faster computation, and deeper insights into the quantum world. As we continue our exploration of quantum information theory, we will delve deeper into the intricacies of quantum algorithms, quantum communication, and the profound impact of quantum computing on our technological landscape. Quantum information theory is not only a theoretical framework; it is a powerful tool that promises to reshape the way we process, transmit, and understand information in the quantum age.*

To delve into the phenomenon of entanglement, we must embark on a journey that takes us deep into the heart of quantum mechanics, where the strange and counterintuitive behavior of particles challenges our classical understanding of the world. Entanglement is a fundamental aspect of quantum physics that Einstein famously described as "spooky action at a distance," and it is a phenomenon that continues to captivate and intrigue physicists and researchers to this day. At its core, entanglement involves a unique connection between two or more particles, such that the state of one particle becomes linked to the state of another, regardless of the physical distance that separates them. This connection defies classical notions of separability and independence, as particles that are entangled exhibit correlated behavior that cannot be explained by classical physics. The history of entanglement dates back to the early days of quantum mechanics, with the groundbreaking work of physicists like Albert Einstein, Niels Bohr, and Erwin Schrödinger. Einstein, Podolsky, and Rosen (EPR) formulated a famous thought experiment in 1935, known as the EPR paradox, to highlight what they saw as a potential flaw in the quantum description of reality. In their scenario, two particles, initially in an entangled state, are sent in opposite directions to remote locations, far from each other. EPR argued that if one could measure a property of one particle, such as its position, with great precision, it would instantaneously determine the position of the other particle, regardless of the distance separating them. This would seem to violate the principle of locality, which states that physical effects cannot propagate faster than the speed of light. Bohr and others countered EPR's argument, asserting that the entangled particles did not have definite properties until measured, and their measurement outcomes were inherently random. In essence, the act of measurement "collapsed" the quantum state into one of its possible outcomes, and this collapse was not influenced by the measurement of the distant particle. This debate laid the foundation for our understanding of entanglement and the role of measurements in quantum mechanics. Schrödinger, in response to EPR's paradox, introduced the term "entanglement" and provided a formal description of this phenomenon. He used the example of two entangled particles, such as electrons, in a quantum state known as a singlet state, where the combined spin of the particles was always zero. This meant that if one particle had an "up" spin when measured, the other would have a "down" spin, and vice versa, regardless of the distance between them. Schrödinger's work highlighted the non-local and correlated nature of entanglement, which could not be explained by classical physics. Entanglement has since become a central concept in quantum mechanics, forming the basis for various quantum phenomena and applications. One of the most famous experiments that demonstrated entanglement in a practical setting is the Aspect experiment, conducted by Alain Aspect in the 1980s. Aspect's experiments involved measuring the polarization of entangled photons, particles of light, which exhibited correlated behavior even when separated by significant distances. The results of these experiments confirmed the non-local nature of entanglement and provided strong evidence in favor of quantum mechanics' predictions. The phenomenon of entanglement also plays a crucial role in the development of quantum technologies, such as quantum cryptography and quantum computing. Quantum key distribution (QKD) protocols, like the BB84 protocol, rely on the principles of entanglement to enable secure communication. In QKD, two parties share entangled particles, and any attempt to intercept or eavesdrop on the communication would disrupt the entangled state, alerting the parties to potential security breaches. This property ensures the confidentiality and security of quantum communication channels. Entanglement is also a valuable resource in quantum computing, where it enables the creation of quantum circuits that exploit the correlated behavior of qubits to perform complex computations. Quantum algorithms like Grover's algorithm and quantum teleportation rely on entanglement to achieve tasks that classical computers cannot replicate efficiently. Grover's algorithm accelerates unstructured search problems, while quantum teleportation enables the transmission of quantum information over long distances without physically moving particles. The concept of entanglement extends beyond the realm of photons and electrons, as it can manifest in various physical systems, including atoms, ions, and even macroscopic objects. Recent experiments have demonstrated entanglement between massive particles, such as buckyballs (large carbon molecules) and mechanical oscillators. These experiments showcase the versatility and ubiquity of entanglement in the quantum world. The phenomenon of entanglement raises intriguing questions about the nature of reality and the interconnectedness of particles.

It challenges our classical intuitions and forces us to confront the non-locality inherent in quantum mechanics. Entanglement has also been the subject of philosophical debates, with some physicists pondering whether it implies the existence of a deeper, hidden reality or a "spooky" interconnectedness in the quantum world. While the mysteries of entanglement continue to fuel scientific curiosity and exploration, its practical applications in quantum technologies are already making a significant impact. Entanglement is at the heart of secure quantum communication and quantum computing, promising to revolutionize fields like cryptography, information processing, and materials science. As we delve deeper into the phenomenon of entanglement, we will explore its implications, applications, and ongoing research efforts to harness its power for the benefit of science and technology. Entanglement serves as a reminder that the quantum world is a realm of astonishing complexity and wonder, where the boundaries of classical physics blur, and new possibilities emerge.

Chapter 8: Quantum Gravity and the Universe

To embark on the quest for a quantum theory of gravity, we must venture into the depths of theoretical physics, where the fundamental forces and particles that govern the universe are unraveled and redefined. The story of this search begins with the realization that our current understanding of gravity, described by Albert Einstein's general theory of relativity, stands in contrast to the framework of quantum mechanics that successfully describes the behavior of the other three fundamental forces: electromagnetism and the strong and weak nuclear forces. General relativity provides a beautiful and accurate description of gravity as the curvature of spacetime caused by massive objects, but it exists in a classical realm, separate from the quantum world of particles and fields. The incompatibility between general relativity and quantum mechanics gives rise to the quest for a quantum theory of gravity, a theoretical framework that unifies these two pillars of modern physics. One of the earliest attempts to reconcile gravity with quantum mechanics was made by Werner Heisenberg in the 1930s when he proposed a theory called "matrix theory." Heisenberg's matrix theory aimed to quantize spacetime itself, treating it as a discrete set of points rather than a continuous manifold. While this approach was pioneering, it did not lead to a complete quantum theory of gravity and was eventually supplanted by other approaches. Another notable figure in the early quest for a quantum theory of gravity was Paul Dirac, who attempted to formulate a quantum version of general relativity. Dirac's work laid the groundwork for later developments in quantum gravity, but it did not result in a fully realized theory. The search for a quantum theory of gravity gained momentum in the 1960s and 1970s with the emergence of string theory. String theory postulates that the fundamental building blocks of the universe are not point particles but tiny, vibrating strings. These strings can exist in multiple dimensions, challenging our conventional understanding of spacetime. String theory has the potential to provide a unified framework for all fundamental forces, including gravity, and it offers a path toward quantum gravity. Supergravity, an extension of string theory that incorporates supersymmetry, further contributed to the development of quantum gravity theories. However, string theory and supergravity are mathematically complex and have raised numerous questions and challenges that remain unresolved. Loop quantum gravity is another prominent approach to quantum gravity that emerged in the late 20th century. This theory, developed by theorists such as Carlo Rovelli and Lee Smolin, attempts to quantize gravity by discretizing spacetime into a network of interconnected loops. In loop quantum gravity, the fabric of spacetime is granular, and fundamental entities known as "spin networks" describe the geometry of the universe. This approach offers a different perspective on the nature of spacetime, distinct from the continuous curvature of general relativity. Loop *quantum gravity has shown promise in addressing some of the issues that arise when trying to reconcile gravity and quantum mechanics. Quantum field theory in curved spacetime, pioneered by Stephen Hawking and others, is yet another approach to quantum gravity. This framework combines quantum field theory, which successfully describes the behavior of particles in flat spacetime, with the effects of gravity. Quantum field theory in curved spacetime has provided valuable insights into phenomena like Hawking radiation, where black holes emit radiation due to quantum effects near their event horizons. However, this approach does not offer a complete quantum theory of gravity and encounters difficulties in dealing with the fundamental nature of spacetime itself. Other approaches, such as causal set theory and causal dynamical triangulations, have also been explored in the quest for a quantum theory of gravity. Each of these approaches presents unique insights and challenges, but none has yet achieved a widely accepted and fully realized theory of quantum gravity. The search for a quantum theory of gravity is not merely a theoretical pursuit; it has profound implications for our understanding of the universe at its most fundamental level. A successful quantum theory of gravity would provide insights into the nature of spacetime, the behavior of matter and energy at extreme scales, and the origin and fate of the cosmos. It would shed light on the enigmatic nature of black holes, the singularity at the center of the Big Bang, and the ultimate structure of the universe. Furthermore, a quantum theory of gravity could offer new possibilities for the unification of all fundamental forces, providing a coherent framework for physics beyond the Standard Model. Despite the challenges and complexities involved, the pursuit of a quantum theory of gravity continues to be a driving force in theoretical physics. Researchers around the world are exploring novel approaches, conducting experiments, and pushing the boundaries of our knowledge. In addition to the fundamental insights it promises, quantum gravity may have practical implications for technologies like space travel and the study of extreme astrophysical phenomena. While the search for a quantum theory of gravity remains ongoing, it exemplifies the spirit of scientific inquiry and the relentless pursuit of understanding the fundamental laws of the universe. It reminds us that even the most profound mysteries can be approached with creativity, perseverance, and collaboration among brilliant minds in the field of theoretical physics. As we venture further into this quest for a unified theory of gravity and quantum mechanics, we explore the frontiers of human knowledge and our capacity to unravel the deepest mysteries of the cosmos. In our exploration of the intersection between quantum mechanics and gravity, we uncover a realm of profound cosmological implications that challenge our understanding of the universe's origin, evolution, and fate. At the heart of this inquiry lies the quest for a quantum theory of gravity, a framework that harmonizes the principles of quantum mechanics with the all-encompassing influence of gravity as described by Albert Einstein's general theory of relativity. This pursuit has ignited a revolution in theoretical physics, as it seeks to bridge the gap between the quantum world of particles and fields and the vast cosmic scales governed by gravity. One of the most intriguing and perplexing aspects of the universe is its origin, often referred to as the "Big Bang." According to the prevailing cosmological model, the universe began as a singularity—an infinitely dense and hot point— approximately 13.8 billion years ago. At this moment of cosmic birth, the laws of physics, including gravity, underwent dramatic changes that are not fully understood within the framework of classical general relativity. Herein lies the first cosmological implication of quantum gravity: it offers the potential to describe the universe's earliest moments with greater precision and coherence. Classical general relativity, while remarkably successful in describing the universe's large-scale structure and dynamics, encounters a singularity at the very heart of the Big Bang. This singularity is a mathematical point where the laws of physics break down, and our current understanding of gravity ceases to apply. Quantum gravity theories, such as loop quantum gravity and string theory, offer alternatives that may resolve this singularity, providing a consistent description of the universe's birth and evolution. In these theories, the fabric of spacetime undergoes quantum effects near the singularity, preventing the occurrence of a true singularity and replacing it with a "bounce" or a transition from a previous contracting phase. The ability to explore the universe's origin with the tools of quantum gravity has profound implications for our understanding of cosmic evolution. Another striking aspect of the universe is its expansion, which was first discovered by the astronomer Edwin Hubble in the 1920s. Hubble's observations revealed that galaxies were moving away from each other, leading to the formulation of Hubble's law, which describes the relationship between the redshift of light from distant galaxies and their recessional velocities. The expansion of the universe presents the next cosmological implication of quantum gravity. Within the framework of classical general relativity, the expansion suggests that the universe is currently in an accelerating phase, driven by an enigmatic entity known as dark energy. While this explanation fits the observational data, it raises questions about the nature of dark energy and its influence on cosmic evolution. Quantum gravity theories may shed light on these questions by providing insights into the fundamental properties of spacetime and the underlying dynamics of the universe. Some quantum gravity models propose modifications to the laws of gravity on cosmological scales, potentially altering the predictions for the expansion of the universe and the nature of dark energy. Understanding these modifications could lead to a deeper understanding of the cosmic acceleration and the fate of the universe. In addition to its implications for the universe's birth and expansion, quantum gravity also plays a crucial role in the study of black holes. Black holes are regions of spacetime where gravity is so intense that nothing, not even light, can escape their gravitational pull. Within the framework of classical general relativity, black holes* possess event horizons, surfaces beyond which nothing can return. However, quantum mechanics introduces a new layer of complexity in the form of Hawking radiation, a phenomenon predicted by Stephen Hawking in 1974. Hawking radiation arises from quantum effects near the event horizon of a black hole, leading to the emission of particles and a gradual loss of mass. This radiation has profound implications for the fate of black holes, as it suggests that they can slowly "evaporate" over time. Quantum gravity theories seek to reconcile this quantum process with the classical concept of black holes and may provide a more complete understanding of their behavior. Furthermore, the interplay between quantum mechanics and gravity gives rise to the black hole information paradox, a long-standing puzzle in theoretical physics. The paradox arises when considering what happens to the information of matter and energy that falls into a black hole. According to the principles of quantum mechanics, information should never be lost, but the classical picture of black holes suggests that information is irretrievably trapped behind the event horizon. Quantum gravity theories, such as string theory, offer potential resolutions to this paradox, suggesting that information may be encoded on the event horizon or in subtle quantum correlations. Exploring the resolution of the black hole information paradox not only deepens our understanding of black holes but also has implications for the broader principles of quantum mechanics and gravity. Beyond the confines of our local universe, the quest for a quantum theory of gravity extends to the study of cosmic phenomena like *dark matter and dark energy. Dark matter is an enigmatic form of matter that does not emit, absorb, or interact with light or electromagnetic forces, yet its gravitational effects are observed on cosmic scales. Quantum gravity theories may provide insights into the nature of dark matter, its interactions with gravity, and its influence on the large-scale structure of the universe. Similarly, the mysterious dark energy, which drives the accelerated expansion of the universe, remains a profound cosmological puzzle. Quantum gravity could illuminate the underlying dynamics of dark energy and its connection to the fundamental laws of physics. The cosmological implications of quantum gravity extend to the very fabric of spacetime itself. In classical general relativity, spacetime is a continuous and smooth manifold, described by smooth curves and surfaces. Quantum gravity challenges this classical view, suggesting that spacetime may have a discrete and granular structure at the smallest scales. This concept of spacetime "foam" or "grains" has profound implications for the behavior of particles and fields on microscopic scales. Quantum gravity theories propose that spacetime fluctuations at the Planck scale, the smallest scale in the universe, may give rise to observable effects, potentially leading to departures from classical physics. The study of cosmic microwave background radiation and high-energy cosmic rays may provide experimental evidence for the granularity of spacetime, offering a glimpse into the quantum nature of the cosmos. As we delve deeper into the realms of quantum gravity, we confront questions that stretch the boundaries of human knowledge. We explore the nature of the universe's birth, the mysteries of cosmic expansion, the enigma of black holes, and the identity of dark matter and dark energy. The pursuit of a quantum theory of gravity is not merely an academic endeavor; it is a quest to unravel the fundamental principles that govern the cosmos. It is a journey that challenges our understanding of the universe and holds the promise of profound discoveries that could reshape our perception of reality itself. In the ongoing exploration of these cosmological implications, we continue to push the frontiers of science, embracing the inherent curiosity that drives humanity's pursuit of knowledge.*

Chapter 9: Bridging the Gap: String Theory and Quantum Computing

In our exploration of the frontiers of theoretical physics and cutting-edge technology, we encounter a fascinating intersection where the abstract world of string theory meets the practical realm of quantum computing. This convergence of two seemingly disparate fields holds the potential to reshape our understanding of the universe and revolutionize the way we process information. String theory, a theoretical framework that describes the fundamental building blocks of the universe as tiny vibrating strings, offers a novel perspective on the nature of reality. It transcends the traditional paradigm of point-like particles and introduces the concept of extra dimensions beyond the familiar three of space and one of time. Within the framework of string theory, the vibrational modes of these strings correspond to different particles, allowing for the unification of all fundamental forces, including gravity. Quantum computing, on the other hand, harnesses the principles of quantum mechanics to perform computations that would be infeasible for classical computers. At its core are quantum bits, or qubits, which can exist in multiple states simultaneously, thanks to the phenomenon of superposition. This enables quantum computers to explore multiple solutions to complex problems in parallel, potentially providing exponential speedup for specific tasks. The convergence of string theory and quantum computing presents an intriguing opportunity to address some of the most profound questions in theoretical physics. One such question is the reconciliation of gravity with the principles of quantum mechanics. In classical general relativity, gravity is described as the curvature of spacetime by massive objects, while quantum mechanics governs the behavior of particles and fields on microscopic scales. String theory offers a potential bridge between these two domains, as it introduces the notion of strings vibrating in spacetime, implying a fundamental connection between gravity and quantum mechanics. Quantum computing, with its ability to simulate quantum systems and explore the behavior of particles in gravitational fields, provides a powerful tool for investigating this connection. Researchers are exploring the use of quantum algorithms to study string theory's predictions and simulate the behavior of strings in various spacetime backgrounds. These simulations offer insights into the quantum properties of black holes, the early universe, and the fundamental structure of spacetime itself. One area of focus is the study of black holes, enigmatic cosmic objects where gravity is so intense that not even light can escape. String theory provides a framework for understanding the microstructure of black holes, including the hypothetical "fuzzball" model, which posits that black holes are composed of highly excited strings. Quantum computing can simulate the behavior of these strings within the black hole, shedding light on their properties and the quantum effects near the event horizon. These *simulations may help resolve long-standing puzzles, such as the black hole information paradox, which concerns the fate of information that falls into a black hole. The interplay between string theory and quantum computing also extends to the study of cosmology and the early universe. String theory introduces the concept of compactified extra dimensions, which could have played a role in the universe's evolution during its early moments. Quantum computing allows researchers to explore the dynamics of these extra dimensions and their implications for the cosmic microwave background radiation and the large-scale structure of the universe. By simulating the quantum behavior of strings in various cosmological scenarios, quantum computers can contribute to our understanding of the universe's origin and evolution. Furthermore, quantum computing holds promise for addressing challenges in string theory calculations. String theory involves intricate mathematical calculations and simulations that can be computationally demanding. Quantum computers, with their potential for exponential speedup, may significantly accelerate these calculations, enabling researchers to explore string theory's predictions more efficiently. Quantum algorithms designed for string theory applications can optimize computations related to string scattering amplitudes, string interactions, and the properties of string states. These advancements can lead to deeper insights into the consistency and viability of string theory as a fundamental theory of the universe. The connection between string theory and quantum computing also extends to the study of quantum gravity itself. While string theory represents a promising approach to a quantum theory of gravity, it poses complex mathematical and theoretical challenges. Quantum computing offers a new avenue for exploring these challenges and testing the predictions of string theory in the context of quantum gravity. Researchers are developing quantum algorithms to investigate the behavior of strings in extreme gravitational environments, such as the vicinity of a black hole or during the early moments of the universe. These simulations can provide valuable insights into the quantum properties of spacetime, the nature of gravity, and the fundamental structure of the cosmos. In addition to its theoretical implications, the convergence of string theory and quantum computing has practical applications in the field of quantum information science. Quantum algorithms inspired by string theory can have broader uses beyond fundamental physics. For example, quantum computers may be employed to optimize complex systems, solve optimization problems, or simulate quantum materials with unique properties. These applications harness the mathematical and computational tools developed in the context of string theory to address real-world challenges. As we navigate this intriguing intersection of theoretical physics and quantum technology, we are poised to uncover new frontiers in our understanding of the universe. The marriage of string theory and quantum computing represents a marriage of abstract theory and practical experimentation. It offers the potential to answer some of the universe's most profound questions, from the nature of gravity and the behavior of black holes to the origins of our cosmos. Simultaneously, it promises advancements in quantum computing, enabling us to tackle complex problems in diverse fields. This convergence exemplifies the power of interdisciplinary exploration, where the boundaries between theory and experimentation blur, and innovation flourishes at the nexus of ideas. As we venture further into this uncharted territory, we embrace the spirit of scientific inquiry and the quest to unlock the secrets of the cosmos through the marriage of string theory and quantum computing. As we navigate the intricate landscape where quantum mechanics and string theory converge, we encounter a realm rich with potential synergies and challenges that shape the future of theoretical physics and technological innovation. At the heart of this exploration is the profound quest to unify the fundamental forces of nature, bridging the gap between the quantum world of particles and fields and the cosmic scales governed by gravity. One remarkable synergy that emerges from the interplay of quantum mechanics and string theory is the potential for a comprehensive theory of quantum gravity. String theory, with its vibrating strings that describe particles and its extended dimensions of spacetime, offers a unique framework that naturally incorporates gravity into the quantum realm. This unification could provide answers to some of the most pressing questions in physics, from the nature of black holes and the singularity at the universe's birth to the ultimate structure of spacetime. Quantum mechanics, on the other hand, contributes its principles of superposition, entanglement, and quantum states, which could deepen our understanding of string theory's predictions and provide new tools for exploring the quantum properties of spacetime itself. The synergy between quantum mechanics and string theory extends to the realm of cosmology, where we seek to comprehend the universe's origin and evolution. String theory's extra dimensions and compactification scenarios offer insights into the early moments of the universe, potentially shedding light on the cosmic microwave background radiation and the large-scale structure of the cosmos. Quantum computing, with its ability to simulate quantum systems and gravitational effects, can accelerate our exploration of these scenarios, enabling us to probe the quantum behavior of strings during the universe's formative stages. Furthermore, the synergy between quantum mechanics and string theory holds promise for addressing longstanding puzzles, such as the black hole information paradox. Quantum mechanics' principles suggest that information cannot be lost, yet classical black holes seem to violate this principle by trapping information behind their event horizons. String theory's microstructure of black holes, combined with quantum computing's simulation capabilities, may offer resolutions to this paradox, revealing the fate of information within black holes. While the potential synergies between quantum mechanics and string theory are exciting, they are not without their challenges. One significant challenge lies in the mathematical complexity of both fields. String theory involves intricate calculations related to string interactions, scattering amplitudes, and the properties of string states. Quantum computing can potentially accelerate these calculations, but developing quantum algorithms for these purposes is a nontrivial task. Quantum algorithm design requires expertise in both quantum mechanics and string theory, demanding interdisciplinary collaboration to unlock the full potential of this synergy. Furthermore, the theoretical landscape of string theory encompasses various formulations, such as perturbative and non-perturbative approaches, heterotic strings, and M-theory, each posing unique challenges and opportunities for quantum computing applications. Navigating this diverse landscape requires a comprehensive understanding of string theory's intricacies and the development of tailored quantum algorithms for specific string theoretic calculations. Another challenge arises from the inherent limitations of quantum hardware. Quantum computers are still in their infancy, and building fault-tolerant quantum devices capable of handling complex string theory calculations remains a formidable task. Noise, decoherence, and error correction are significant obstacles that must be addressed to realize the full potential of quantum computing in string theory applications. Moreover, the computational power of quantum devices may be limited by the availability of qubits and their connectivity, which can impact the scale and accuracy of simulations. Despite these challenges, the convergence of quantum mechanics and string theory represents a tantalizing frontier that demands exploration. To harness the full potential of this synergy, collaboration between theorists, experimentalists, and quantum computing experts is essential. Interdisciplinary research teams can work together to develop quantum algorithms, design quantum hardware optimized for string theory simulations, and conduct experiments that validate string theory predictions. The challenges encountered along this journey offer opportunities for innovation, pushing the boundaries of quantum technology and theoretical physics. Moreover, the convergence of quantum mechanics and string theory extends beyond the realm of fundamental physics. The mathematical and computational tools developed in this context can have broader applications in areas such as materials science, optimization, cryptography, and data analysis. Quantum algorithms inspired by string theory may find practical use in solving complex problems, optimizing quantum materials, or simulating quantum systems for advanced technological applications. As we navigate the intricate landscape where quantum mechanics and string theory converge, we are on the cusp of transformative discoveries that could reshape our understanding of the universe and propel quantum computing into new frontiers. The potential synergies between these two fields offer a path to unlocking profound insights into the fundamental laws of nature and addressing some of the most perplexing questions in physics. While challenges lie ahead, the spirit of exploration and collaboration drives us forward, beckoning us to embrace the opportunities that arise at the intersection of quantum mechanics and string theory. In this journey of discovery, we embark on a quest to unite the forces of quantum mechanics and gravity, unravel the mysteries of the cosmos, and pave the way for a quantum future where the boundaries of knowledge and technology are continually pushed to new horizons.*

Chapter 10: Future Horizons: Advances in Quantum Physics

In the ever-evolving landscape of quantum physics, we find a dynamic field marked by continuous exploration and discovery. Emerging trends in quantum physics reflect the relentless pursuit of deeper understanding and the quest for practical applications that could reshape technology and our perception of reality. One prominent trend that has captured the attention of physicists and technologists alike is the rapid advancement of quantum computing. Quantum computers, with their ability to perform complex calculations exponentially faster than classical computers for certain tasks, hold the promise of revolutionizing fields such as cryptography, materials science, and optimization. The race to develop scalable and fault-tolerant quantum hardware has intensified, with leading tech companies and research institutions vying for supremacy in the nascent quantum computing industry. As quantum computing matures, it opens up possibilities for tackling previously intractable problems, such as simulating quantum systems, factoring large numbers efficiently, and optimizing complex processes. Another notable trend is the exploration of quantum materials and their unique properties. Quantum materials exhibit exotic phenomena, such as high-temperature superconductivity, topological insulators, and quantum spin liquids, which challenge our conventional understanding of matter and its behavior. Researchers are delving into the fundamental physics of these materials, aiming to harness their properties for applications in energy storage, quantum computing, and next-generation electronics. The emergence of quantum materials as a thriving research area reflects the growing recognition of their potential to usher in a new era of technology. Quantum communication, a field focused on secure and unbreakable communication using quantum properties like entanglement and quantum key distribution, continues to gain momentum. Quantum cryptography protocols, such as the BBM92 and E91 protocols, offer secure communication channels immune to eavesdropping, making them invaluable in fields like financial transactions and government communications. Quantum networks, which enable the distribution of entangled particles across large distances, are being developed as the backbone of future quantum internet infrastructure. These quantum communication technologies are poised to play a pivotal role in safeguarding sensitive information and reshaping the cybersecurity landscape. Quantum sensing and metrology represent another burgeoning trend with applications in precision measurement and fundamental physics experiments. Quantum sensors, leveraging the principles of quantum mechanics, offer unprecedented levels of sensitivity and accuracy in fields like gravitational wave detection, magnetic resonance imaging (MRI), and environmental monitoring. These sensors have the potential to revolutionize medical diagnostics, mineral exploration, and geophysics, while *also advancing our understanding of the natural world. The field of quantum optics continues to push boundaries, as researchers explore quantum phenomena involving the interaction of light and matter. Quantum optics experiments, such as quantum teleportation and quantum teleportation across long distances, challenge our intuition about the nature of particles and the fundamental principles governing the quantum world. These experiments not only deepen our understanding of quantum mechanics but also hold promise for future quantum communication and quantum computing applications. Quantum simulations, a versatile trend in quantum physics, enable the emulation of complex quantum systems that are difficult to study with classical computers. Researchers are using quantum simulators to investigate the behavior of molecules, materials, and quantum phases, shedding light on quantum chemistry, condensed matter physics, and beyond. These simulations have the potential to accelerate drug discovery, optimize materials for energy applications, and explore the behavior of matter under extreme conditions. The trend of quantum machine learning, an interdisciplinary field at the intersection of quantum computing and artificial intelligence, is on the rise. Quantum machine learning algorithms promise to outperform classical machine learning methods by exploiting quantum properties like superposition and entanglement. These algorithms have applications in data analysis, pattern recognition, optimization, and autonomous systems, with the potential to revolutionize industries ranging from healthcare to finance. Quantum gravity and the quest for a complete theory of quantum gravity remain a compelling trend in theoretical physics. String theory, loop quantum gravity, and other approaches seek to reconcile the principles of quantum mechanics with the nature of gravity, addressing the mysteries of black holes, the singularity at the universe's birth, and the fundamental structure of spacetime. These efforts represent a grand intellectual challenge that could reshape our understanding of the fundamental forces of the universe. The burgeoning field of quantum biology explores the role of quantum phenomena in biological processes. Researchers investigate how quantum effects such as tunneling, entanglement, and coherence may play a role in processes like photosynthesis, enzyme reactions, and sensory perception. Quantum biology offers insights into the fundamental mechanisms of life and has potential applications in drug design and biotechnology. In the realm of quantum technology, the trend of miniaturization and integration of quantum devices is evident. Researchers are developing compact quantum sensors, quantum memories, and quantum processors that can fit on semiconductor chips. These miniaturized quantum devices open the door to practical applications in portable quantum technologies, quantum-enhanced sensors, and quantum communication systems. The concept of quantum supremacy, a milestone indicating when quantum computers outperform classical computers for a specific task, has gained prominence. Achieving quantum supremacy represents a significant trend, as it signals the maturation of quantum* computing and its readiness to tackle real-world problems. Quantum algorithms are being designed and tested to demonstrate this pivotal moment in the development of quantum technology. The trend of interdisciplinary collaboration is pivotal in advancing quantum physics. Researchers from diverse fields, including physics, computer science, chemistry, and engineering, are coming together to tackle complex challenges and harness the potential of quantum technologies. Interdisciplinary approaches facilitate breakthroughs in quantum computing, quantum materials, quantum communication, and other quantum domains. The trend of global collaboration in quantum research is also evident, with countries and institutions forming international partnerships to accelerate progress in quantum science and technology. Quantum research centers, laboratories, and consortia are fostering the exchange of knowledge, expertise, and resources to propel quantum advancements on a global scale. In summary, the ever-evolving landscape of quantum physics is marked by emerging trends that encompass quantum computing, quantum materials, quantum communication, quantum sensing, quantum optics, quantum simulations, quantum machine learning, quantum gravity, quantum biology, and quantum technology. These trends reflect a vibrant field where scientific curiosity converges with technological innovation, offering profound insights into the nature of the quantum world and the potential to transform industries and society. As researchers continue to explore these trends, they pave the way for a quantum *future where the boundaries of knowledge and technology are continually expanded, and the quantum realm becomes an integral part of our scientific and technological landscape.*

As we embark on a journey through the ever-evolving landscape of quantum physics, it is imperative to consider the anticipated breakthroughs and the lingering mysteries that continue to captivate the minds of physicists and researchers around the world. One of the most eagerly anticipated breakthroughs is the achievement of quantum supremacy, a momentous milestone in the field of quantum computing. Quantum supremacy refers to the point at which a quantum computer surpasses the computational capabilities of even the most powerful classical supercomputers for a specific task. This achievement would signify not only the maturation of quantum technology but also the potential for quantum computers to revolutionize fields such as cryptography, materials science, and optimization. Quantum supremacy would open the door to tackling complex problems that were previously deemed intractable, ushering in a new era of computational capabilities. Another area of great anticipation is the development of practical quantum applications. Quantum technologies, such as quantum communication and quantum sensing, are rapidly advancing and hold the promise of practical, real-world use. Secure quantum communication networks, impervious to eavesdropping, could transform the way we transmit sensitive information, making quantum cryptography a reality. Quantum sensors, with their unparalleled sensitivity and precision, have applications in fields like medicine, environmental monitoring, and fundamental physics experiments. The realization of these quantum applications could have profound implications for various industries and scientific endeavors. In the realm of quantum materials, the discovery of new exotic phases and the engineering of quantum states are eagerly anticipated. Quantum materials exhibit unconventional behaviors, such as high-temperature superconductivity, topological insulators, and quantum spin liquids. Understanding and harnessing these properties could lead to groundbreaking advances in energy storage, electronics, and quantum computing. The quest for room-temperature superconductors, which would revolutionize power transmission and transportation, remains a tantalizing goal for researchers. Additionally, the engineering of topological quantum states may open the door to robust and fault-tolerant quantum computers. One of the grand challenges in quantum physics is the quest for a complete theory of quantum gravity. The reconciliation of general relativity, which describes gravity at cosmological scales, with quantum mechanics, which governs the behavior of particles on the smallest scales, remains an unresolved puzzle. String theory, loop quantum gravity, and other approaches seek to provide a unified framework that accounts for both quantum and gravitational effects. The discovery of experimental evidence or theoretical breakthroughs that bridge this gap would mark a monumental achievement in our understanding of the fundamental forces of the universe. Quantum entanglement, often described as "spooky action at a distance" by Albert Einstein, remains a topic of fascination and intrigue. Understanding the underlying mechanisms of entanglement and harnessing it for practical purposes continue to be active areas of research. Quantum entanglement has the potential to enable secure quantum communication and teleportation, but its full implications and applications are still being explored. The mysteries of black holes, gravitational singularities, and the behavior of matter under extreme conditions remain at the forefront of theoretical physics. Breakthroughs in these areas could reshape our understanding of the universe and provide insights into the nature of spacetime itself. For example, the resolution of the black hole information paradox, which concerns the fate of information that falls into a black hole, is a longstanding puzzle that continues to challenge researchers. The development of quantum algorithms for simulating black holes and exploring their quantum properties holds promise for unraveling this enigma. The quest to uncover the true nature of dark matter and dark energy, which together constitute the majority of the universe's mass-energy content, remains an unsolved mystery. Physicists are actively searching for direct evidence of dark matter particles and probing the nature of dark energy through cosmological observations. Breakthroughs in understanding these cosmic mysteries could revolutionize our comprehension of the cosmos. Quantum biology, a nascent field that explores the role of quantum phenomena in biological processes, holds the potential to unveil the quantum secrets of life itself. Researchers are investigating how quantum effects, such as tunneling, entanglement, and coherence, may play a role in processes like photosynthesis, enzyme reactions, and sensory perception. Unlocking the quantum mechanisms of life could have implications for drug discovery and biotechnology. The quest for quantum-safe cryptography, in anticipation of the threat posed by future quantum computers to classical encryption methods, is a pressing concern. Researchers are developing quantum-resistant encryption schemes and exploring the use of quantum cryptography to secure data in a post-quantum era. The realization of practical quantum-resistant cryptography is essential to safeguarding sensitive information in the future. Quantum supremacy, practical quantum applications, quantum gravity, quantum entanglement, black hole mysteries, dark matter, quantum biology, and quantum-safe cryptography represent some of the most eagerly anticipated breakthroughs and unanswered questions in the realm of quantum physics. These frontiers of knowledge and technology beckon researchers to push the boundaries of our understanding and explore the profound mysteries of the quantum world. As we venture deeper into the quantum realm, we remain poised on the cusp of transformative discoveries that have the potential to reshape our world and expand the horizons of human knowledge.

BOOK 3
QUANTUM PHYSICS DEMYSTIFIED
FROM NOVICE TO QUANTUM EXPERT

ROB BOTWRIGHT

Chapter 1: Introduction to Quantum Physics

In the early 20th century, a scientific revolution was underway, giving birth to a new branch of physics known as quantum mechanics. This revolutionary theory emerged as a response to the limitations of classical physics in explaining the behavior of matter and energy at the atomic and subatomic scales. At the heart of this transformation was the realization that the classical laws of physics, which had successfully described the motion of planets and the behavior of everyday objects, were insufficient to explain the strange and counterintuitive phenomena observed in the microscopic world. The seeds of quantum physics were sown in the late 19th century with the advent of quantum theory, a precursor to the full-fledged quantum mechanics that would later emerge. One of the pivotal figures in this early development was Max Planck, a German physicist who is often considered the father of quantum theory. In 1900, Planck introduced the concept of quantization to explain the spectral distribution of blackbody radiation. He proposed that energy could only be emitted or absorbed in discrete, quantized units, which he called "quanta." This revolutionary idea challenged the classical notion of continuous energy and laid the groundwork for the quantum revolution. Planck's quantization of energy was a departure from the classical physics of the time, and it was met with skepticism by some of his contemporaries. However, it provided an elegant solution to the problem of blackbody radiation and offered a glimpse into the quantum nature of the universe. Albert Einstein, another luminary of the era, further advanced quantum theory in 1905 with his explanation of the photoelectric effect. Einstein proposed that light consists of discrete packets of energy called photons, which interact with matter in quantized fashion. His work not only confirmed the existence of quanta but also earned him the Nobel Prize in Physics in 1921. As quantum theory continued to evolve, it became clear that the classical deterministic view of the universe, where every particle's position and momentum could be precisely predicted, was fundamentally flawed at the quantum level. Werner Heisenberg's uncertainty principle, formulated in 1927, asserted that the more precisely we know a particle's position, the less precisely we can know its momentum, and vice versa. This inherent uncertainty in quantum measurements shattered the classical notion of determinism, introducing an element of unpredictability into the quantum world. Niels Bohr, a Danish physicist, developed the Copenhagen interpretation of quantum mechanics, which emphasized the role of probability and wave functions in describing quantum systems. Bohr's model of the atom, with electrons occupying discrete energy levels, provided a framework for understanding atomic spectra and chemical behavior. The Bohr model laid the foundation for the emerging field of quantum chemistry, where quantum mechanics played a central role in explaining chemical bonding and molecular structure. In parallel with these developments, Erwin

Schrödinger formulated wave mechanics in 1926, introducing the concept of wave functions to describe the behavior of particles. Schrödinger's wave equation, a fundamental equation of quantum mechanics, provided a mathematical framework for understanding wave-particle duality and the probabilistic nature of quantum systems. Wave functions represented the probability amplitudes of finding particles in different states, and their solutions revealed the quantized energy levels of quantum systems. This wave-particle duality, where particles exhibit both wave-like and particle-like behavior, was a fundamental departure from classical physics. The advent of quantum mechanics brought about a shift in our understanding of physical reality. It revealed that particles, such as electrons and photons, do not have well-defined trajectories but exist as probability distributions described by wave functions. This probabilistic nature of quantum systems challenged our classical intuition and led to famous thought experiments, such as Schrödinger's cat paradox, which illustrated the paradoxical and counterintuitive aspects of quantum theory. One of the profound consequences of quantum mechanics was the concept of quantum entanglement.

In 1935, Albert Einstein, Boris Podolsky, and Nathan Rosen published a paper presenting the EPR paradox, which highlighted the non-local and correlated nature of entangled particles. Einstein famously referred to this phenomenon as "spooky action at a distance." Entanglement occurs when two or more particles become correlated in such a way that the measurement of one particle's property instantaneously affects the state of the other, regardless of the distance separating them. This phenomenon challenged classical notions of locality and causality and posed deep philosophical questions about the nature of reality. As quantum mechanics continued to mature, it found practical applications in technology and industry. The development of quantum mechanics paved the way for advancements in fields such as quantum optics, quantum electronics, and quantum computing. Quantum optics explored the behavior of light and matter at the quantum level, leading to the development of lasers, masers, and other quantum-based technologies. In the realm of quantum electronics, the invention of the transistor, based on the principles of quantum tunneling, revolutionized the field of electronics and paved the way for modern computing. The advent of quantum computing, with the potential to perform complex calculations at speeds unimaginable to classical computers, represents one of the most exciting frontiers in quantum physics. In recent years, quantum computers from companies like IBM, Google, and others have demonstrated quantum supremacy by solving specific tasks faster than classical supercomputers. These developments have sparked enthusiasm for quantum computing's potential to revolutionize fields such as cryptography, materials science, and optimization. Quantum physics also played a crucial role in the development of quantum field theory, a framework that combines quantum mechanics with special relativity to describe the behavior of particles and fields in a unified manner. Quantum field theory provided a comprehensive understanding of the electromagnetic, weak, and strong nuclear forces, leading to the development of the standard model of particle physics. This model successfully explained the behavior of elementary particles and their interactions, culminating in the discovery of the Higgs boson in 2012. The standard model, based on quantum field theory, stands as one of the most successful theories in the history of physics. Yet, despite its successes, the standard model leaves unanswered questions about the nature of dark matter, dark energy, and the unification of all fundamental forces. The birth of quantum physics was a transformative moment in the history of science. It challenged classical notions of determinism, introduced the probabilistic nature of the quantum world, and led to the development of revolutionary technologies. Quantum mechanics, with its wave functions, uncertainty principle, and entanglement, continues to captivate the imagination of physicists and researchers, offering profound insights into the fundamental nature of the universe. As we delve deeper into the quantum realm, we are confronted with both the marvels and mysteries of this extraordinary branch of physics, where the boundaries of human understanding are continually pushed, and the quest for answers   continues.

The early decades of the 20th century were marked by a series of groundbreaking experiments and discoveries that laid the foundation for the development of quantum physics. One of the earliest experiments that challenged classical physics was the photoelectric effect, which was first observed by Heinrich Hertz in 1887. However, it was Albert Einstein's explanation of the photoelectric effect in 1905 that revolutionized our understanding of the interaction between light and matter. Einstein proposed that light consists of discrete packets of energy called photons, and when these photons strike a material, they can eject electrons from the surface. The energy of the ejected electrons depends on the frequency of the light, not its intensity, which was contrary to classical predictions. This discovery provided strong evidence for the quantization of energy and the particle-like behavior of light. In 1913, Niels Bohr introduced the Bohr model of the hydrogen atom, which incorporated the quantization of energy levels. Bohr's model successfully explained the spectral lines of hydrogen, but it also introduced the concept of quantum jumps, where electrons transition between discrete energy levels by absorbing or emitting photons. This model was a significant departure from classical mechanics and set the stage for the development of quantum theory. Another pivotal experiment that challenged classical physics was the discovery of the Compton effect in 1923 by Arthur Compton. Compton observed that when X-rays scattered off electrons, they shifted to longer wavelengths, indicating that they behaved as particles with discrete energies. This phenomenon provided further support for the particle-wave duality of light and matter. Around the same time,

Louis de Broglie proposed the idea of wave-particle duality, suggesting that particles like electrons could exhibit both wave-like and particle-like properties. De Broglie's hypothesis was confirmed in 1927 when Clinton Davisson and Lester Germer observed electron diffraction patterns, similar to those of X-rays, when electrons were directed at a crystalline nickel target. This experimental evidence supported de Broglie's idea that particles have associated wave properties. In 1926, Erwin Schrödinger formulated wave mechanics, a mathematical framework that described the behavior of quantum particles using wave functions. Schrödinger's wave equation, a fundamental equation of quantum mechanics, allowed scientists to calculate the probability distributions of particles in various quantum states. This marked a significant shift from the classical notion of deterministic trajectories to a probabilistic description of particle behavior. Wave mechanics provided a unified framework for understanding the wave-particle duality of matter and energy. In 1927, Werner Heisenberg developed matrix mechanics, another formalism for quantum mechanics that used matrices to represent physical observables. Heisenberg's uncertainty principle, formulated the same year, introduced the concept that the more precisely one knows a particle's position, the less precisely one can know its momentum, and vice versa. This inherent uncertainty in measurements was a fundamental departure from classical determinism and introduced an element of unpredictability into quantum physics. In 1928, Paul Dirac combined the principles of wave mechanics and matrix mechanics to create a more comprehensive quantum mechanics known as quantum field theory. Quantum field theory treated particles as excitations of quantum fields and successfully explained the behavior of electrons and other subatomic particles. The development of quantum mechanics revolutionized our understanding of atomic and subatomic physics, but it also gave rise to a host of new phenomena and paradoxes. One of the most puzzling aspects of quantum physics was the phenomenon of quantum entanglement. In 1935, Albert Einstein, Boris Podolsky, and Nathan Rosen published a paper presenting the EPR paradox, which highlighted the non-local and correlated nature of entangled particles. Einstein famously referred to this phenomenon as "spooky action at a distance." Entanglement occurs when two or more particles become correlated in such a way that the measurement of one particle's property instantaneously affects the state of the other, regardless of the distance separating them. This phenomenon challenged classical notions of locality and causality and posed deep philosophical questions about the nature of reality. In 1964, physicist John Bell formulated Bell's theorem, which provided a way to experimentally test the predictions of quantum entanglement. Subsequent experiments, such as those by Alain Aspect in the 1980s, confirmed that the predictions of quantum mechanics held, and that entanglement was indeed a real and non-classical phenomenon. Quantum entanglement remains one of the most enigmatic and fascinating aspects of quantum physics. The development of quantum physics also led to the emergence of quantum statistics, a branch of physics that describes the behavior of particles with integer or half-integer spin, such as electrons, protons, and neutrons. Quantum statistics introduced the concept of Fermi-Dirac statistics for fermions, which obey the Pauli exclusion principle and cannot occupy the same quantum state simultaneously. Bosons, on the other hand, follow Bose-Einstein statistics and can occupy the same quantum state, leading to phenomena like Bose-Einstein condensates at ultra-low temperatures. These statistical principles played a crucial role in understanding the behavior of matter at low temperatures and in the development of technologies such as superconductors and superfluids. Quantum physics also had a profound impact on the field of chemistry. The quantization of energy levels in atoms and molecules explained the discrete spectral lines observed in atomic and molecular spectra. This understanding led to the development of quantum chemistry, which used quantum mechanics to predict the electronic structure and properties of molecules. Quantum chemistry has played a vital role in fields like materials science, drug discovery, and chemical engineering. The development of quantum mechanics also had implications for the study of nuclear physics. The quantization of energy levels in atomic nuclei explained the stability and behavior of isotopes, and it provided insights into the processes of nuclear fusion and fission. This understanding laid the groundwork for nuclear physics and had implications for the development of nuclear energy. In 1928, physicist Paul

Dirac introduced the Dirac equation, which combined quantum mechanics with special relativity to describe the behavior of relativistic electrons. The Dirac equation predicted the existence of antimatter, a mirror-image counterpart to ordinary matter with opposite charge. This prediction was confirmed with the discovery of the positron, the first antiparticle, by Carl D. Anderson in 1932. The existence of antimatter had profound implications for our understanding of particle physics and cosmology. The development of quantum field theory in the mid-20th century led to the formulation of the standard model of particle physics, which described the behavior of elementary particles and their interactions through the electromagnetic, weak, and strong nuclear forces. The standard model successfully explained the behavior of particles and predicted the existence of particles like the W and Z bosons, which were later discovered. In 2012, the discovery of the Higgs boson at the Large Hadron Collider provided experimental confirmation of the Higgs field, which gives particles mass and is a cornerstone of the standard model. However, despite its successes, the standard model leaves unanswered questions about the nature of dark matter, dark energy, and the unification of all fundamental forces. The early quantum experiments and discoveries of the 20th century transformed our understanding of the physical world. They ushered in a new era of physics, where the probabilistic and wave-particle nature of matter and energy became central to our understanding of the universe. These foundational discoveries paved the way for the development of quantum mechanics, quantum field theory, and the standard model of particle physics, providing insights into the behavior of matter and the fundamental forces that govern the cosmos.

Chapter 2: The Quantum Revolution

In the late 19th and early 20th centuries, the foundations of classical physics, which had successfully described the behavior of the physical world for centuries, began to show signs of strain. Classical physics, based on Newtonian mechanics and Maxwell's equations of electromagnetism, had provided a robust framework for understanding the motion of objects and the behavior of electromagnetic waves. It had successfully explained the orbits of planets, the motion of projectiles, and the propagation of light and electricity. However, as scientists delved deeper into the microscopic realm and explored the behavior of particles at the atomic and subatomic levels, classical physics revealed its limitations. One of the first challenges to classical physics came from the study of blackbody radiation, the electromagnetic radiation emitted by a perfectly absorbing and radiating body. Classical electromagnetic theory predicted that the intensity of blackbody radiation should increase without bound as the frequency of radiation increased, a prediction known as the "ultraviolet catastrophe." This divergence from experimental data was a clear indication that classical physics was inadequate in describing the behavior of radiation at the atomic scale. Max Planck's groundbreaking work in 1900 introduced the concept of quantization, where energy could only be emitted or absorbed in discrete units or "quanta." Planck's quantization of energy successfully explained the spectral distribution of blackbody radiation and marked the birth of quantum theory. It challenged the classical notion of continuous energy and introduced the idea that energy levels are quantized in the atomic world. This was a departure from classical physics, where energy was treated as continuous and infinitely divisible. Albert Einstein further advanced the understanding of quantization by explaining the photoelectric effect in 1905. He proposed that light consists of discrete packets of energy called photons, each carrying a specific amount of energy proportional to its frequency. Einstein's work not only explained the photoelectric effect but also provided compelling evidence for the particle-like behavior of light. This was a departure from classical wave theory, which had treated light as a continuous wave. In 1913, Niels Bohr introduced the Bohr model of the hydrogen atom, incorporating quantization of energy levels. Bohr's model successfully explained the spectral lines of hydrogen but introduced the concept of quantum jumps, where electrons transition between discrete energy levels by absorbing or emitting photons. This model was another step away from classical physics, where electrons were expected to follow continuous trajectories. As quantum theory continued to develop, it became clear that classical physics was unable to explain the behavior of particles at the atomic and subatomic scales. The classical determinism, where the position and momentum of a particle could be precisely predicted, broke down in the quantum world. Werner Heisenberg's uncertainty principle, formulated in 1927, stated that the more precisely one knows a particle's position, the less precisely one can know its momentum, and vice versa. This inherent uncertainty in quantum measurements challenged the classical notion of deterministic predictability. Niels Bohr further elaborated on the probabilistic nature of quantum mechanics with the Copenhagen interpretation, which emphasized the role of wave functions and probability distributions in describing quantum systems. Wave functions represented the probability amplitudes of finding particles in various states, and their square magnitudes represented the probability densities. This was a profound departure from classical physics, where particles were treated as definite objects with precise properties. Erwin Schrödinger, in 1926, formulated wave mechanics, a mathematical framework that described quantum behavior using wave functions. Schrödinger's wave equation became a fundamental equation of quantum mechanics, allowing scientists to calculate the probability distributions of particles in different quantum states. Wave mechanics provided a unified framework for understanding the wave-particle duality and probabilistic nature of quantum systems. Wave functions represented the quantum state of a system, and their evolution over time was governed by the Schrödinger equation. Another significant departure from classical physics was the concept of quantum entanglement. In 1935, Albert Einstein, Boris Podolsky, and Nathan Rosen published a paper presenting the EPR paradox, which highlighted the non-local and correlated nature of entangled particles. Einstein famously referred to this phenomenon as "spooky action at a distance." Entanglement occurs when two or more particles become correlated in such a way that the measurement of one particle's property instantaneously affects the state of the other, regardless of the distance separating them. This phenomenon challenged classical notions of locality and causality, raising deep philosophical questions about the nature of reality. As quantum mechanics continued to evolve, it found practical applications in technology and industry. The development of quantum mechanics paved the way for advancements in fields such as quantum optics, quantum electronics, and quantum computing. Quantum optics explored the behavior of light and matter at the quantum level, leading to the development of lasers, masers, and other quantum-based technologies. In the realm of quantum electronics, the invention of the transistor, based on the principles of quantum tunneling, revolutionized the field of electronics and paved the way for modern computing. The advent of quantum computing, with the potential to perform complex calculations at speeds unimaginable to classical computers, represents one of the most exciting frontiers in quantum physics. In recent years, quantum computers from companies like IBM, Google, and others have demonstrated quantum supremacy by solving specific tasks faster than classical supercomputers. These developments have sparked enthusiasm for quantum computing's potential to revolutionize fields such as cryptography, materials science, and *optimization. Quantum physics also played a crucial role in the development of quantum field theory, a framework that combines quantum mechanics with special relativity to describe the behavior of particles and fields in a unified manner. Quantum field theory provided a comprehensive understanding of the electromagnetic, weak, and strong nuclear forces, leading to the development of the standard model of particle physics. This model successfully explained the behavior of elementary particles and their interactions, culminating in the discovery of the Higgs boson in 2012. The standard model, based on quantum field theory, stands as one of the most successful theories in the history of physics. Yet, despite its successes, the standard model leaves unanswered questions about the nature of dark matter, dark energy, and the unification of all fundamental forces. The breakdown of classical physics in the face of quantum phenomena marked a profound shift in our understanding of the physical world. It introduced concepts such as quantization, wave-particle duality, and quantum entanglement that challenged classical determinism and determinacy. As we delve deeper into the quantum realm, we are confronted with both the marvels and mysteries of this extraordinary branch of physics, where the boundaries of human understanding are continually pushed, and the quest for answers continues.*

Chapter 3: Understanding Quantum States

Quantum mechanics, with its foundation in wave functions and state vectors, revolutionized our understanding of the fundamental building blocks of the universe. These mathematical constructs play a central role in describing the quantum state of a physical system. In quantum mechanics, the state of a system is represented by a state vector, often denoted as $|\Psi\rangle$. This state vector contains all the information about the system's quantum properties, including the positions, momenta, and spins of its constituent particles. The state vector evolves in time according to the Schrödinger equation, a fundamental equation of quantum mechanics. The Schrödinger equation describes how the quantum state changes over time as the system's Hamiltonian operator acts on it. The Hamiltonian operator represents the total energy of the system, including kinetic and potential energies. Solving the Schrödinger equation allows us to predict the future quantum state of the system or understand its past evolution. The concept of a state vector is closely related to wave functions, which provide a detailed description of a quantum system's spatial and temporal properties. Wave functions are often denoted as $\Psi(x)$, where x represents the position of the particle in space. The magnitude squared of the wave function, $|\Psi(x)|^2$, gives the probability density of finding the particle at a particular position x. This interpretation of the wave *function's square magnitude as a probability density is a fundamental aspect of quantum mechanics, distinguishing it from classical physics. Wave functions are complex-valued functions, meaning they have both real and imaginary parts. The complex nature of wave functions allows them to encode information about the phase of a quantum system, which is essential for understanding interference and other quantum phenomena. Wave functions can describe single particles or entire quantum systems, such as atoms, molecules, or even larger objects. In the case of a single particle, the wave function $\Psi(x)$ provides a complete description of the particle's quantum state. For multi-particle systems, the wave function becomes a multi-dimensional function that depends on the positions of all the particles. The behavior of wave functions is governed by the principles of superposition and linearity. Superposition allows multiple quantum states to be combined to form new states. For example, if $|\Psi_1\rangle$ and $|\Psi_2\rangle$ are two possible quantum states, their superposition $|\Psi\rangle = a|\Psi_1\rangle + b|\Psi_2\rangle$, where a and b are complex coefficients, represents a new quantum state that is a linear combination of the two original states. Superposition is a fundamental property of quantum systems, leading to phenomena such as interference and entanglement. Interference occurs when two or more quantum states overlap and interfere constructively or destructively, affecting the probabilities of measurement outcomes. Entanglement, on the other hand, describes the correlation between the quantum states of two or more particles, even when they are* separated by large distances. This non-local correlation is a hallmark of quantum entanglement and has been the subject of much research and fascination. The concept of wave functions and state vectors is not limited to position or spatial coordinates. In quantum mechanics, there are also wave functions that describe other observable properties, such as momentum, spin, and angular momentum. These wave functions provide a comprehensive description of a quantum system's properties, allowing us to make predictions about measurement outcomes. The uncertainty principle, formulated by Werner Heisenberg, imposes limits on the precision with which certain pairs of complementary properties, such as position and momentum, can be simultaneously known. This principle arises from the mathematical properties of wave functions and the inherent probabilistic nature of quantum measurements. The mathematical relationship between position and momentum wave functions is described by the Fourier transform, which allows us to understand the trade-off between knowledge of position and momentum. Wave functions can be represented in different mathematical formalisms, such as the position representation or the momentum representation. In the position representation, the wave function $\Psi(x)$ gives the probability amplitude for finding a particle at position x. In the momentum representation, the wave function $\Phi(p)$ describes the probability amplitude for the particle to have momentum p. These representations are related through Fourier transforms, which provide a mathematical bridge between the two descriptions. The *mathematical formalism of wave functions and state vectors is a powerful tool for solving quantum mechanical problems. By representing physical systems in terms of state vectors and applying the Schrödinger equation, physicists can make predictions about the behavior of quantum systems, from the behavior of electrons in atoms to the behavior of particles in particle accelerators. Wave functions and state vectors are essential in understanding the behavior of quantum systems in diverse fields, including quantum chemistry, solid-state physics, and quantum information science. Quantum computers, for example, rely on manipulating quantum state vectors to perform calculations that would be infeasible for classical computers. Understanding the mathematics and principles of wave functions and state vectors is crucial for anyone seeking to explore the fascinating world of quantum mechanics and its many applications in science and technology.*

Top of Form

Chapter 4: The Principles of Quantum Mechanics

The Schrödinger equation, a cornerstone of quantum mechanics, governs the evolution of wave functions over time. Erwin Schrödinger formulated this equation in 1926, providing a powerful framework for describing the quantum behavior of physical systems. The Schrödinger equation is fundamentally a wave equation, analogous to the classical wave equations used to describe phenomena like sound and light. However, unlike classical waves, quantum wave functions describe the probabilistic distribution of particles in a quantum system. The time-dependent Schrödinger equation describes how a quantum system's wave function changes with time. It is written as follows: $i\hbar\ \partial\Psi/\partial t = H\Psi$, where $\hbar$ (pronounced "h-bar") is the reduced Planck's constant, $\partial\Psi/\partial t$ represents the partial derivative of the wave function with respect to time, H is the Hamiltonian operator, and Ψ is the wave function. The Hamiltonian operator, denoted as H, represents the total energy operator of the quantum system. It includes the kinetic and potential energy operators for all particles in the system. Solving the time-dependent Schrödinger equation allows us to determine how the wave function evolves over time, providing insights into the quantum behavior of particles. The time-independent Schrödinger equation, on the other hand, deals with stationary states where the energy of the system does not change with time. It is written as: $H\Psi = $

$E\Psi$, where E represents the total energy of the system and Ψ is the wave function. Solving the time-independent Schrödinger equation yields the energy eigenstates and corresponding wave functions for the quantum system. These energy eigenstates are the quantized energy levels that a quantum system can occupy. The solutions to the Schrödinger equation are wave functions, denoted as Ψ, that describe the probability amplitudes of finding particles in different quantum states. These wave functions are complex-valued and contain both real and imaginary components. The square magnitude of the wave function, $|\Psi|^2$, represents the probability density of finding a particle in a particular quantum state. The interpretation of $|\Psi|^2$ as a probability density is a fundamental aspect of quantum mechanics, distinguishing it from classical physics. The behavior of wave functions is governed by the principles of superposition and linearity. Superposition allows multiple quantum states to be combined to form new states. For example, if $|\Psi 1\rangle$ and $|\Psi 2\rangle$ are two possible quantum states, their superposition $|\Psi\rangle = a|\Psi 1\rangle + b|\Psi 2\rangle$ represents a new quantum state that is a linear combination of the two original states. Superposition is a fundamental property of quantum systems, leading to phenomena such as interference and entanglement. Interference occurs when two or more quantum states overlap and interfere constructively or destructively, affecting the probabilities of measurement outcomes. Entanglement, on the other hand, describes the correlation between the quantum states of two or more *particles, even when they are separated by large distances. This non-local correlation is a hallmark of quantum entanglement and has been the subject of much research and fascination. Wave functions can describe the quantum state of a single particle or an entire quantum system, such as an atom, molecule, or larger object. For a single particle, the wave function provides a complete description of the particle's quantum properties, including its position, momentum, and spin. For multi-particle systems, the wave function becomes a multi-dimensional function that depends on the positions and properties of all the particles. Wave functions can be represented in different mathematical formalisms, such as the position representation or the momentum representation. In the position representation, the wave function $\Psi(x)$ describes the probability amplitude for finding a particle at a particular position x. In the momentum representation, the wave function $\Phi(p)$ describes the probability amplitude for the particle to have momentum p. These representations are related through mathematical transforms, such as the Fourier transform, which provides a mathematical bridge between the two descriptions. The mathematical formalism of wave functions and the Schrödinger equation is a powerful tool for solving quantum mechanical problems. By applying the Schrödinger equation to physical systems, physicists can make predictions about the behavior of quantum systems, from the behavior of electrons in atoms to the behavior of particles in particle accelerators. Wave functions and the Schrödinger* *equation are essential in understanding the behavior of quantum systems in diverse fields, including quantum chemistry, solid-state physics, and quantum information science. Quantum computers, for example, rely on manipulating quantum states and applying the Schrödinger equation to perform calculations that would be infeasible for classical computers. Understanding the mathematics and principles of wave functions and the Schrödinger equation is crucial for anyone seeking to explore the fascinating world of quantum mechanics and its many applications in science and technology.*

Chapter 5: Quantum Measurement and Uncertainty

Quantum measurement is a fundamental aspect of quantum mechanics, and it plays a central role in understanding the behavior of quantum systems. In quantum mechanics, the wave function describes the quantum state of a system, representing all possible outcomes of measurements. When a quantum measurement is performed, it reveals a specific value of the measured property, such as position, momentum, or spin. This process is known as wavefunction collapse or quantum state reduction. Wavefunction collapse is a concept that has intrigued and puzzled physicists since the early days of quantum mechanics. The collapse of the wave function occurs because measurements in quantum mechanics are inherently probabilistic. Before a measurement, the wave function represents a superposition of all possible measurement outcomes, each with an associated probability amplitude. These probabilities are calculated using the squared magnitudes of the wave function components. When a measurement is made, one specific outcome is observed, and the wave function "collapses" into the corresponding eigenstate of the measured property. This means that after the measurement, the quantum system is in a definite state, no longer described by a superposition of possibilities. The exact mechanism of wavefunction collapse and its interpretation have been the subject of much debate and various interpretations within the field of quantum mechanics. One interpretation, often associated with the Copenhagen interpretation, posits that wavefunction collapse is a fundamental and irreducible aspect of quantum measurement. According to this view, the act of measurement itself causes the collapse of the wave function, and the choice of outcome is probabilistic. Another interpretation, known as the Many-Worlds interpretation, proposes that wavefunction collapse doesn't occur, but rather, all possible outcomes of a measurement exist in separate, non-communicating branches of the quantum universe. In this view, the observer experiences one outcome but is unaware of the other branches. The debate over wavefunction collapse and its interpretation highlights the deep philosophical and foundational questions raised by quantum mechanics. One important consequence of wavefunction collapse is the Heisenberg uncertainty principle, which limits the precision with which certain pairs of complementary properties, such as position and momentum, can be simultaneously known. This limitation arises because, in a state of superposition before measurement, the uncertainty in one property is related to the spread of possible outcomes in the other property. When a measurement is made, the wave function collapses, and the uncertainty in one property becomes sharply defined, while the uncertainty in the complementary property increases. This intrinsic uncertainty is a fundamental feature of quantum mechanics and sets it apart from classical physics. The concept of wavefunction collapse has practical implications in the laboratory, as it affects the precision and predictability of quantum measurements. In experiments with quantum systems, scientists must carefully consider the effects of wavefunction collapse and account for the inherent uncertainty in measurements. The phenomenon of wavefunction collapse also challenges our classical intuitions about the nature of reality. In the macroscopic world, we are accustomed to deterministic outcomes and definite properties of objects. However, in the quantum realm, wavefunctions describe a world of inherent probabilities and potentialities, where the act of observation itself can change the state of a system. The measurement problem in quantum mechanics, which deals with the interpretation of wavefunction collapse, remains an active area of research and philosophical inquiry. Many experiments have confirmed the probabilistic nature of quantum measurements and the wavefunction collapse phenomenon, but the underlying mechanisms and their philosophical implications continue to be subjects of exploration and debate. As our understanding of quantum mechanics deepens, so too does our appreciation of the profound and puzzling aspects of the quantum world. The concept of wavefunction collapse challenges our classical intuitions and invites us to grapple with the fundamental nature of reality at the quantum level. While the interpretation of quantum mechanics remains a topic of philosophical inquiry, the mathematical framework of the theory has proven remarkably successful in describing and predicting the behavior of particles at the smallest scales. Quantum mechanics has led to countless technological

advancements and continues to drive scientific discovery, even as its mysteries continue to captivate and inspire physicists and philosophers alike.
Top of Form

Chapter 6: Quantum Entanglement

Quantum entanglement is one of the most intriguing and puzzling phenomena in the realm of quantum mechanics. It is a phenomenon where two or more particles become correlated in such a way that the properties of one particle are instantly connected to the properties of another, even when they are separated by vast distances. This bizarre behavior was famously described by Albert Einstein as "spooky action at a distance," and it challenges our classical intuitions about the nature of reality. The phenomenon of entanglement was first introduced in a 1935 paper by Albert Einstein, Boris Podolsky, and Nathan Rosen, commonly known as the EPR paper. In this paper, the authors presented a thought experiment to highlight what they perceived as a paradox in quantum mechanics.

They argued that if quantum mechanics were a complete theory, it would allow for instantaneous correlations between distant particles, which seemed incompatible with the theory of relativity. However, at the time, it was just a thought experiment, and the phenomenon of entanglement was not yet experimentally confirmed. It wasn't until 1964 that physicist John Bell formulated a set of mathematical inequalities, now known as Bell's inequalities, that could be tested in experiments to determine whether entanglement was a real phenomenon. Bell's inequalities provided a way to distinguish between the predictions of quantum mechanics and those of classical theories that assumed hidden variables governing the behavior of particles. Experiments conducted in the following decades, starting with the work of John Clauser, Stuart Freedman, and others, consistently violated Bell's inequalities, providing strong experimental evidence in favor of entanglement and disproving the idea of hidden variables. These experiments showed that entanglement was not just a mathematical curiosity but a real and fundamental aspect of quantum physics. Entanglement occurs when two or more particles, such as electrons, photons, or atoms, interact in a way that their quantum states become intertwined. These particles are described by a joint quantum state that cannot be separated into individual states, even when they are far apart. One of the key features of entanglement is that the properties of the entangled particles are no longer independent. For example, if two electrons are entangled in terms of their spins, measuring the spin of one electron instantly determines the spin of the other, regardless of the distance separating them.

This instantaneous correlation between entangled particles is known as quantum non-locality and has been confirmed in numerous experiments. Einstein, Podolsky, and Rosen's thought experiment, which aimed to highlight the paradox of entanglement, challenged the completeness of quantum mechanics. They argued that if the theory were complete, measuring one entangled particle's properties should provide information about the other particle's properties, even when the two particles were spacelike separated, meaning no signal could travel between them at the speed of light. This seemed to violate the principles of relativity, which state that no information or influence can travel faster than the speed of light.

However, quantum mechanics doesn't rely on the transmission of information between entangled particles. Instead, it predicts correlations between their properties that are established at the moment of entanglement and maintained regardless of the spatial separation. This property of entanglement doesn't violate relativity because it doesn't involve the transfer of information. Entanglement is a fundamental aspect of quantum mechanics and has been experimentally observed in various physical systems. One common example is the entanglement of photons, the particles of light. In a quantum optics experiment, two entangled photons can be created, and measurements of one photon's properties instantly affect the state of the other, even when they are sent in opposite directions over large distances. This phenomenon has practical applications in quantum communication, where entangled particles are used to create secure communication channels that are theoretically immune to eavesdropping. Another example of entanglement is seen in the entangled states of electron spins in certain materials. These entangled electron pairs can exhibit correlated behaviors, such as the emission of correlated photons, which is essential in technologies like quantum dots and quantum information processing. Entanglement is also at the heart of quantum teleportation, a process that allows the transfer of quantum information from one location to another without physically moving the particles themselves. Quantum teleportation relies on the entanglement between particles to transmit quantum states instantaneously over long distances.

The phenomenon of entanglement has deep implications for our understanding of the nature of reality. It challenges the classical notion of separability, suggesting that particles can exist in a state of quantum entanglement where their properties are interdependent, even when they are widely separated. This has led to debates about the philosophical implications of quantum entanglement, such as the nature of physical reality, determinism, and the role of consciousness in quantum measurement. While entanglement remains a perplexing and mysterious aspect of quantum mechanics, it is also a powerful resource with practical applications in quantum technology. Quantum computers, for example, leverage the properties of entanglement to perform certain types of calculations at speeds unattainable by classical computers. Quantum cryptography utilizes entangled particles to create secure communication systems that are theoretically immune to hacking. Understanding and harnessing the phenomenon of entanglement is a central focus of quantum research, and it promises to *revolutionize fields ranging from computing to communication to materials science. As we continue to explore the phenomenon of quantum entanglement, we uncover not only the strange and fascinating properties of the quantum world but also the potential for transformative advances in technology and our understanding of the universe.*

Chapter 7: Quantum Operators and Observables

Operators are fundamental mathematical constructs in quantum mechanics, playing a central role in describing and understanding the behavior of quantum systems. They are used to represent physical observables, such as position, momentum, energy, and angular momentum, in a quantum context. Operators provide a bridge between the mathematical formalism of quantum mechanics and the physical reality of the quantum world. In quantum mechanics, observables are associated with Hermitian operators, which are a special class of linear operators. A Hermitian operator is one that is equal to its adjoint or conjugate transpose, ensuring that its eigenvalues (possible measurement outcomes) are real. One of the most well-known Hermitian operators is the Hamiltonian operator, denoted as H, which represents the total energy of a quantum system. The eigenvalues of the Hamiltonian operator correspond to the allowed energy levels of the system, and its eigenvectors represent the corresponding quantum states. Operators also allow us to define transformations that act on quantum states, changing them in various ways. For example, the position operator, denoted as X, can be used to describe the position of a particle in one dimension. When applied to a quantum state, it returns the position of the particle as an observable quantity. The momentum operator, denoted as P, represents the momentum of a particle and is related to the rate of change of the position operator with respect to time. The commutation relationship between position and momentum operators, [X, P] = XP - PX, is a fundamental expression in quantum mechanics and leads to the Heisenberg uncertainty principle. This principle states that the product of the uncertainties in position and momentum is bounded by ℏ/2, where ℏ is the reduced Planck's constant. Operators also enable the description of physical transformations in quantum mechanics, such as time evolution. The time evolution operator, denoted as U(t), is responsible for how quantum states change over time. It is a fundamental concept in quantum mechanics, allowing us to predict how quantum systems evolve from one state to another as time progresses. The Schrödinger equation, which governs the time evolution of quantum states, is intimately connected to the time evolution operator. Operators can be combined through mathematical operations, such as addition, subtraction, multiplication, and division, to create new operators. For example, the kinetic energy operator for a particle with mass m can be expressed as (P^2)/(2m), where P is the momentum operator. This operator combines the momentum operator with mathematical operations to represent kinetic energy as an observable. Operators also play a crucial role in the description of quantum states as vectors in a complex vector space, known as Hilbert space. In this framework, quantum states are represented as state vectors, and operators act on these vectors to produce new states. The inner product of two state vectors is used to calculate probabilities and determine the overlap between different quantum states. The inner product is an essential mathematical tool for calculating transition probabilities and understanding the quantum measurement process. Quantum mechanics provides a rich mathematical framework for dealing with the uncertainty and probabilistic nature of the quantum world. Operators and their corresponding eigenvalues and eigenvectors are central to this framework, allowing us to make predictions about the outcomes of measurements and understand the behavior of quantum systems. The mathematical formalism of operators is both powerful and elegant, providing a unifying language for describing the behavior of particles at the quantum level. Quantum mechanics also introduces the concept of operators that do not commute, meaning their order of application matters. The non-commutativity of operators is a fundamental aspect of quantum mechanics and leads to the Heisenberg uncertainty principle, which sets limits on our ability to simultaneously know certain pairs of observables, such as position and momentum. Operators can also be used to represent physical transformations in quantum mechanics, such as rotations in space. For example, the angular momentum operator, denoted as L, represents the angular momentum of a particle, and its eigenvalues correspond to quantized values of angular momentum. The eigenstates of the angular momentum operator are associated with specific orientations in space. Operators are not limited to single-particle systems but can also describe the behavior of composite systems composed *of multiple particles. In these cases, operators act on the combined Hilbert space of all particles, allowing us to analyze and predict the behavior of complex quantum systems. Operators are also essential in quantum chemistry, where they are used to describe the behavior of electrons in atoms and molecules. Operators like the electronic Hamiltonian operator are employed to solve the Schrödinger equation for multi-electron systems, providing insights into chemical bonding and molecular properties. Quantum mechanics has been incredibly successful in explaining and predicting the behavior of particles and physical systems at the quantum level. Operators and their algebraic properties are at the heart of this success, providing the mathematical foundation for understanding the quantum world. While the mathematics of operators can be challenging, their role in quantum mechanics is essential for scientists and researchers seeking to explore and harness the fascinating and sometimes counterintuitive behavior of particles in the quantum realm.*

Chapter 8: Quantum Mechanics in Practice

Solving quantum problems and applying quantum principles has become an area of increasing interest and importance in the fields of science and technology. Quantum mechanics, with its unique mathematical framework and conceptual foundations, allows us to address a wide range of problems and explore innovative applications. One of the primary challenges in solving quantum problems is dealing with the probabilistic nature of quantum mechanics. Unlike classical physics, where outcomes are deterministically predictable, quantum systems are characterized by inherent uncertainty. This uncertainty is encapsulated by the wave function, which describes a quantum system's probability distribution of possible states. To solve quantum problems, we often employ mathematical tools and techniques that involve the manipulation of wave functions and operators. Quantum algorithms, such as those used in quantum computing, have the potential to revolutionize problem-solving in various domains. Quantum computers leverage the principles of superposition and entanglement to perform calculations that would be infeasible for classical computers. For instance, Shor's algorithm can efficiently factor large numbers, posing a significant challenge to classical encryption methods. Grover's algorithm accelerates database searching, promising speedup for various optimization problems. Quantum simulators, another application of quantum computing, enable researchers to model and understand complex quantum systems, such as the behavior of molecules and materials. Solving problems related to quantum chemistry, for example, is of immense importance for drug discovery and materials design. Quantum algorithms also have implications for machine learning and artificial intelligence, where they can enhance optimization tasks and data analysis. Another application of quantum mechanics is in quantum cryptography, a field that aims to secure communication using the principles of quantum physics. Quantum key distribution (QKD) protocols enable two parties to exchange secret keys with unconditional security. By exploiting the properties of quantum entanglement and the no-cloning theorem, QKD provides a new level of security against eavesdropping. Quantum cryptography has the potential to revolutionize secure communication in an era of increasing cyber threats. Quantum metrology, the science of precise measurement using quantum systems, offers applications in fields like navigation, geodesy, and fundamental constants determination. Quantum-enhanced sensors can achieve unprecedented levels of accuracy and sensitivity, impacting industries ranging from healthcare to environmental monitoring. Quantum communication networks, which utilize entangled particles for secure data transmission, could provide global-scale secure communication infrastructure. These networks may become crucial for protecting sensitive information in an increasingly interconnected world. Quantum cryptography and communication also hold *promise for secure voting systems, ensuring the integrity of elections and decision-making processes. Quantum algorithms have shown potential in optimizing supply chain management, logistics, and transportation, leading to more efficient and sustainable systems. Quantum computing's impact extends to the optimization of financial portfolios and risk assessment in the finance industry. Additionally, quantum simulations have the potential to revolutionize materials science by accelerating the discovery of new materials with desired properties. Quantum mechanics is instrumental in the development of quantum sensors for medical imaging, enabling early disease detection and improved patient care. Quantum technologies can enhance the security of personal identification and financial transactions, reducing the risk of identity theft and fraud. Quantum-enhanced imaging techniques, such as quantum-enhanced microscopy, offer the potential for higher-resolution imaging in fields like biology and medicine. Quantum-enhanced sensors can be used in environmental monitoring to detect pollutants and address climate-related challenges. Quantum cryptography can secure critical infrastructure and protect against cyberattacks on power grids, water systems, and transportation networks. Quantum mechanics has opened new frontiers in materials science, enabling the design of advanced materials with unique properties for various applications. Quantum algorithms have the potential to revolutionize optimization problems in fields like logistics, finance, and transportation. Quantum computing's impact on* cryptography could reshape the way we secure data and protect sensitive information in an increasingly digital world. Quantum communication networks have the potential to provide secure, long-distance communication, making them invaluable in industries like finance, healthcare, and defense. Quantum technologies are driving advancements in medical imaging, enabling early disease detection and personalized medicine. Quantum-enhanced sensors and imaging techniques are essential tools for environmental monitoring and addressing global challenges related to climate change and pollution. Quantum computing's potential to accelerate scientific research and discovery has implications for fields ranging from materials science to drug development. Quantum mechanics has unlocked the potential for quantum sensors and imaging techniques with applications in a wide range of industries and scientific disciplines. Quantum technologies hold the promise of transforming industries, revolutionizing problem-solving, and addressing some of the most pressing challenges facing society. As our understanding of quantum mechanics deepens and our ability to harness quantum principles advances, the impact of quantum technologies on our world is expected to grow exponentially, offering innovative solutions to complex problems and shaping the future of science, technology, and industry.

Chapter 9: Advanced Topics in Quantum Physics

Quantum field theory (QFT) is a profound theoretical framework that unifies quantum mechanics with special relativity. It extends the principles of quantum mechanics to fields, allowing us to describe the behavior of particles as excitations of quantum fields. Relativistic quantum mechanics, on the other hand, combines the principles of special relativity with quantum mechanics to describe particles moving at high speeds. The need for both special relativity and quantum mechanics became apparent with the advent of particle physics, as classical physics failed to explain the behavior of subatomic particles traveling at significant fractions of the speed of light. The development of relativistic quantum mechanics and quantum field theory was a monumental achievement in theoretical physics. One of the fundamental concepts in relativistic quantum mechanics is the energy-momentum relation, which relates the energy and momentum of a particle to its mass and velocity. This relation, given by $E^2 = (pc)^2 + (mc^2)^2$, where E is the energy, p is the momentum, m is the rest mass, and c is the speed of light, is a cornerstone of special relativity. In relativistic quantum mechanics, particles are described by wave functions that satisfy relativistically invariant wave equations, such as the Dirac equation for fermions and the Klein-Gordon equation for bosons. These equations take into account the effects of special relativity, ensuring that quantum mechanics remains consistent at high speeds.

Quantum field theory extends the principles of quantum mechanics to fields, such as the electromagnetic field and the electron field. In QFT, particles are considered as localized excitations of these fields, and the field itself becomes the primary object of study. The quantum field is described by a field operator that creates and annihilates particles. For example, the electromagnetic field is quantized using field operators that create and annihilate photons. Quantum field theory also introduces the concept of quantized angular momentum, known as spin. Particles are classified into two broad categories based on their spin: fermions, which have half-integer spin, and bosons, which have integer spin. Fermions obey the Pauli exclusion principle, which states that no two identical fermions can occupy the same quantum state simultaneously. This principle underlies the structure of matter and the stability of atoms, as it prevents electrons in an atom from occupying the same energy level. Bosons, on the other hand, can occupy the same quantum state, and they mediate the fundamental forces of nature, such as photons for electromagnetism and gluons for the strong nuclear force. Quantum field theory also introduces the concept of virtual particles, which are particles that temporarily exist as intermediate states in quantum interactions. These virtual particles play a crucial role in the quantum description of fundamental forces. For instance, in quantum electrodynamics (QED), the theory of electromagnetic interactions, virtual photons mediate the electromagnetic force between charged particles. Quantum field theory provides a framework for calculating the probability amplitudes of particle interactions and predicting the outcomes of high-energy experiments. These calculations involve perturbation theory, where interactions are treated as small corrections to a free field theory. Feynman diagrams, graphical representations of particle interactions, are used to visualize and calculate these amplitudes. Feynman diagrams are a powerful tool for understanding and calculating complex particle interactions in QFT. Quantum field theory has been highly successful in describing the behavior of particles and fields in the context of particle physics. The Standard Model of particle physics, which incorporates the electromagnetic, weak, and strong nuclear forces, is based on QFT principles and has been experimentally confirmed to a remarkable degree. However, there are still challenges and open questions in the field of quantum field theory. One of the central challenges is the unification of quantum field theories, particularly the unification of the Standard Model with gravity. General relativity, Einstein's theory of gravity, is a classical theory, and attempts to quantize gravity using the principles of QFT have encountered mathematical and conceptual difficulties. This has led to the search for a theory of quantum gravity that can reconcile general relativity with quantum mechanics. String theory is one of the leading candidates for a theory of quantum gravity and aims to provide a unified framework for all fundamental forces, including gravity. In string theory, particles are not point-like but are one-dimensional strings vibrating in multiple dimensions. String theory incorporates the principles of quantum field theory and general relativity, making it a candidate for a theory of everything. Quantum field theory has also played a crucial role in understanding the behavior of particles in extreme environments, such as those found in the early universe or in the vicinity of black holes. These extreme conditions challenge our understanding of quantum field theory and can lead to the creation of exotic particles and phenomena. Quantum field theory continues to be a vibrant area of research, with ongoing efforts to extend and refine its principles and applications. It remains an essential tool for describing the behavior of particles and fields in the quantum realm, from the subatomic scales of particle physics to the cosmological scales of the universe. The interplay between quantum mechanics, special relativity, and quantum field theory has deepened our understanding of the fundamental forces and particles that make up the fabric of the universe, and it continues to inspire new discoveries and insights in the world of physics.

Chapter 10: Becoming a Quantum Expert

Embarking on a journey to explore the depths of quantum physics can be an exhilarating endeavor for those who seek to unravel the mysteries of the quantum world. As you delve deeper into this fascinating field, you will encounter a wide array of topics and concepts that expand your understanding of the fundamental laws governing the universe. One of the first steps on this intellectual voyage is gaining a solid foundation in the principles of quantum mechanics. This foundational knowledge serves as the bedrock upon which you will build your understanding of more advanced topics. Quantum mechanics, with its intricate mathematics and abstract concepts, can initially appear daunting, but with patience and persistence, you will gradually grasp its fundamental principles. The journey begins with the wave-particle duality, a concept that challenges the classical notion of particles having definite positions and velocities. You will learn that particles, such as electrons and photons, exhibit both wave-like and particle-like behavior, depending on how they are observed and measured. Quantum superposition is another fundamental concept that you will encounter. It reveals the astonishing idea that quantum particles can exist in multiple states simultaneously, as long as they are not observed or measured. This concept is exemplified by the famous Schrödinger's cat thought experiment, where a cat inside a closed box can be considered both alive and dead until the box is opened and observed. The uncertainty principle, formulated by Werner Heisenberg, is a cornerstone of quantum mechanics. It asserts that there are inherent limits to our ability to simultaneously measure certain pairs of properties, such as a particle's position and momentum. This principle challenges the classical idea of precise measurement and introduces an element of unpredictability into the behavior of quantum systems. As you continue your journey, you will delve into the fascinating realm of quantum entanglement. This phenomenon occurs when two or more particles become interconnected in such a way that their properties are intertwined, regardless of the distance that separates them. Entangled particles can instantaneously influence each other's states, even when separated by vast distances, a phenomenon referred to as "spooky action at a distance," as famously described by Albert Einstein. To deepen your understanding of quantum mechanics, you will explore the mathematical formalism that underpins the theory. You will become acquainted with wave functions, which describe the quantum state of a system, and operators, which represent physical observables like position, momentum, and angular momentum. The Schrödinger equation, a foundational equation in quantum mechanics, describes how the quantum state of a system evolves over time. Solving this equation allows you to predict the behavior of quantum systems and calculate probabilities of various outcomes. With a solid grasp of the fundamentals, you will embark on a journey into the world of advanced quantum topics. Quantum field theory, a framework that combines quantum mechanics with special relativity, plays a pivotal role in understanding the behavior of particles in high-energy physics. It describes particles as excitations of quantum fields and has been highly successful in predicting and explaining the behavior of subatomic particles. For those with a keen interest in the fundamental forces of nature, quantum field theory provides the tools to explore electromagnetism, the weak and strong nuclear forces, and the elusive quest for a theory of quantum gravity. String theory, a candidate for a theory of everything, takes you into the realm of vibrating strings as the fundamental constituents of the universe. It seeks to unify all known forces and particles within a single, elegant framework, transcending the boundaries of classical physics. String theory introduces extra dimensions beyond the familiar three spatial dimensions and one time dimension, offering a new perspective on the nature of reality. The pursuit of string theory entails a deep dive into complex mathematical structures and the exploration of a landscape of possible string theory scenarios. For those drawn to the intersection of quantum physics and computing, the world of quantum computing awaits. Quantum computers harness the principles of superposition and entanglement to perform calculations that are currently beyond the capabilities of classical computers. This revolutionary technology has the potential to impact fields such as cryptography, optimization, and scientific simulations. Quantum algorithms, such as Shor's algorithm and Grover's algorithm, are at the forefront of quantum computing *research and offer a glimpse into the transformative power of quantum computation. As you delve into the quantum computing landscape, you will encounter quantum gates, qubits, and the intricate choreography of quantum algorithms that manipulate and process quantum information. Quantum computing challenges traditional notions of computation and opens up new horizons for solving complex problems. The study of quantum information theory complements your exploration of quantum computing. It examines how information can be encoded, transmitted, and processed in a quantum mechanical framework. Quantum cryptography, a subset of quantum information theory, promises secure communication through the principles of quantum key distribution. These cryptographic protocols leverage the properties of quantum entanglement to ensure the privacy and integrity of transmitted information. With each step in your journey through the quantum realm, you will encounter a rich tapestry of concepts and applications. Quantum mechanics, quantum field theory, string theory, quantum computing, and quantum information theory form the foundation upon which the edifice of quantum physics stands. Your pursuit of further studies in quantum physics will involve both theoretical exploration and experimental investigation. You may find yourself immersed in cutting-edge research projects, collaborating with physicists from around the world, or working with state-of-the-art laboratory equipment. As you navigate this complex and enthralling landscape, you will contribute to humanity's expanding understanding of the fundamental laws that govern the universe. Whether your passion lies in unraveling the mysteries of the subatomic world, probing the nature of spacetime, or revolutionizing the world of computation, the pursuit of further studies in quantum physics promises a journey of intellectual discovery and scientific advancement. The exploration of quantum phenomena challenges our intuition, stretches the boundaries of our mathematical prowess, and invites us to contemplate the profound implications of a quantum universe. It is a journey that leads to deeper insights, innovative technologies, and a profound appreciation for the elegance and intricacy of the quantum world.*

The advancement of quantum science has been shaped by the remarkable contributions of countless scientists, theorists, and experimentalists who have dedicated their lives to unraveling the mysteries of the quantum world. One of the early pioneers of quantum physics was Max Planck, who, in 1900, introduced the concept of quantization of energy to explain the spectral distribution of blackbody radiation. His work laid the foundation for the development of quantum mechanics and earned him the Nobel Prize in Physics in 1918. Albert Einstein, another towering figure in the history of physics, made significant contributions to quantum theory with his explanation of the photoelectric effect in 1905. His discovery that light behaves as discrete packets of energy, or photons, provided strong evidence for the quantum nature of light. In 1921, Einstein was awarded the Nobel Prize for this groundbreaking work.

Niels Bohr, a Danish physicist, made pivotal contributions to the development of quantum theory, especially in the field of atomic physics. He proposed the Bohr model of the atom in 1913, which incorporated quantized electron energy levels and successfully explained the spectral lines of hydrogen. Bohr's model was instrumental in advancing our understanding of atomic structure. Werner Heisenberg's uncertainty principle, formulated in 1927, revolutionized quantum mechanics by establishing the fundamental limits of precision in measuring certain pairs of complementary properties, such as position and momentum. This principle introduced an element of intrinsic uncertainty into the description of quantum systems. Erwin Schrödinger, an Austrian physicist, developed the Schrödinger equation in 1926, which provided a powerful mathematical framework for describing the behavior of quantum systems. Schrödinger's equation became a cornerstone of quantum mechanics, allowing for the calculation of wave functions and probabilities. Max Born, a German physicist, made a fundamental contribution by interpreting the square of the wave function as the probability density of finding a particle in a particular state. This interpretation laid the groundwork for the probabilistic nature of quantum mechanics. Paul Dirac, a British physicist, made groundbreaking contributions to quantum theory with his formulation of quantum mechanics using linear algebra and the Dirac equation, which described the behavior of relativistic electrons and predicted the existence of antimatter. Dirac's work had a profound impact on the development of quantum field theory. Richard Feynman, an American physicist, introduced the concept of path integrals in quantum mechanics, providing an alternative mathematical framework for describing quantum processes. Feynman diagrams, graphical representations of particle interactions, revolutionized quantum field theory and played a crucial role in calculating particle interaction probabilities. John Bell's theorem, formulated in 1964, demonstrated that certain quantum phenomena, such as entanglement, cannot be explained by classical physics and suggested the existence of non-local correlations between entangled particles. Bell's work inspired experiments that confirmed the predictions of quantum mechanics and challenged classical intuitions about physical reality. In the realm of experimental physics, numerous scientists have made groundbreaking contributions. Otto Stern and Walther Gerlach conducted the famous Stern-Gerlach experiment in 1922, which provided experimental evidence for quantized angular momentum and the quantization of particle spin. Claude Cohen-Tannoudji, Steven Chu, and William D. Phillips were awarded the Nobel Prize in Physics in 1997 for their development of laser cooling techniques, which allowed for the precise manipulation and cooling of atoms, opening new avenues for studying quantum behavior at extremely low temperatures. Alain Aspect's experiments in the 1980s provided strong evidence against local hidden variable theories and supported the concept of quantum entanglement, as predicted by Bell's theorem. David Wineland and Serge Haroche were jointly awarded the Nobel Prize in Physics in 2012 for their groundbreaking work in trapping and manipulating individual ions and photons, paving the way for quantum information processing. The advent of quantum computers, which promise to perform certain calculations exponentially faster than classical computers, has been driven by the contributions of researchers like Peter Shor and Lov Grover. Peter Shor's quantum algorithm for factoring large numbers threatens classical encryption methods and has implications for the security of digital communication. Lov Grover's quantum search algorithm offers a quadratic speedup in searching unsorted databases, revolutionizing information retrieval. The field of quantum information theory, which explores the fundamental limits and possibilities of quantum information processing, has seen significant contributions from Charles Bennett, Gilles Brassard, and John Preskill, among others. Their work has illuminated the potential of quantum cryptography, quantum teleportation, and quantum error correction. Experimentalists such as Alain Aspect, Anton Zeilinger, and Juan Ignacio Cirac have pushed the boundaries of quantum experiments, testing the fundamental principles of quantum mechanics and exploring the phenomena of quantum entanglement and non-locality. The development of quantum technologies, including quantum cryptography, quantum key distribution, and quantum sensors, has benefited from the collective efforts of scientists and engineers who have harnessed the unique properties of quantum systems for practical applications. Quantum teleportation experiments, conducted by teams led by Jian-Wei Pan and Nicolas Gisin, have achieved unprecedented feats of transferring quantum information instantaneously over long distances. The development of quantum hardware, including superconducting qubits, trapped ions, and topological qubits, has been driven by research teams at organizations like IBM, Google, Rigetti, and many others. The quest for building scalable and fault-tolerant quantum computers is a collective endeavor. Quantum science has also extended its reach into the realm of fundamental physics, with experiments designed to test the limits of quantum mechanics, probe the nature of dark matter, and explore the foundations of gravity. The contributions of scientists like Juan Maldacena, who proposed the AdS/CFT correspondence, have led to insights connecting quantum field theory and gravity. Quantum technologies are poised to revolutionize fields beyond physics, including cryptography, healthcare, materials science, and finance. The development of quantum algorithms for optimization and machine learning, led by researchers like Hartmut Neven at Google, holds the promise of transforming industries and solving complex problems more efficiently. The study of quantum materials, driven by physicists like J. C. Séamus Davis and his group, explores the unique properties of materials at the quantum level, offering the potential for breakthroughs in electronics, energy storage, and beyond. In the realm of quantum biology, scientists such as Seth Lloyd and Jim Al-Khalili have investigated the role of quantum phenomena in biological processes, potentially shedding light on the mysteries of consciousness and the origins of life. Quantum science has become an interdisciplinary endeavor, bringing together physicists, engineers, computer scientists, and mathematicians to tackle some of the most challenging questions in science and technology. The collaboration between academia, industry, and government institutions has fueled the rapid progress of quantum research and development. International collaborations and research partnerships have fostered a global community of scientists dedicated to advancing quantum science and technology. Quantum science continues to be a dynamic field, with ongoing research pushing the boundaries of our understanding and capabilities. As we look to the future, the contributions of countless individuals and teams will play a pivotal role in shaping the next era of quantum exploration and innovation. The quest to harness the power of quantum phenomena for practical applications and to deepen our understanding of the fundamental laws of the universe remains a compelling and exciting journey, one that continues to inspire scientists and enthusiasts alike to push the boundaries of what is possible in the quantum realm.

BOOK 4
MASTERING QUANTUM PHYSICS
FROM BASICS TO ADVANCED CONCEPTS

ROB BOTWRIGHT

Chapter 1: Quantum Foundations and Principles

The historical development of quantum physics is a fascinating journey through the evolution of human understanding of the fundamental laws that govern the behavior of the smallest particles in the universe. It is a story that begins at the turn of the 20th century, when the classical physics of Isaac Newton and James Clerk Maxwell faced profound challenges in explaining the behavior of atoms and subatomic particles. At that time, the prevailing view was that the universe operated according to deterministic laws, where the positions and velocities of particles could be precisely predicted if their initial conditions were known. However, a series of experimental discoveries and groundbreaking theories would shatter this classical worldview and pave the way for the birth of quantum physics. One of the earliest hints of the quantum revolution came from Max Planck's work on blackbody radiation in 1900. Planck introduced the concept of quantization of energy, suggesting that energy could only be exchanged in discrete units or quanta, rather than continuously. This radical idea was the first glimpse of a departure from classical physics and would become a cornerstone of quantum theory. In 1905, Albert Einstein made a pivotal contribution by explaining the photoelectric effect, where light falling on a metal surface caused the emission of electrons. Einstein proposed that light consisted of discrete packets of energy, or photons, with each photon carrying a specific amount of energy. This insight provided compelling evidence for the quantization of energy and the particle-like nature of light, challenging the prevailing wave theory of light. Niels Bohr, a Danish physicist, made a significant breakthrough in 1913 with his model of the hydrogen atom. Bohr's model incorporated the idea of quantized electron energy levels, where electrons could only occupy specific orbits around the nucleus. This model successfully explained the discrete spectral lines observed in hydrogen's emission and absorption spectra, marking a pivotal moment in the development of quantum theory. Werner Heisenberg, a German physicist, introduced the uncertainty principle in 1927, which fundamentally changed our understanding of measurement and precision in the quantum world. Heisenberg's principle stated that certain pairs of complementary properties, such as a particle's position and momentum, could not be simultaneously measured with arbitrary precision. This inherent uncertainty challenged the classical notion of deterministic measurements and emphasized the probabilistic nature of quantum mechanics. Erwin Schrödinger, an Austrian physicist, made another profound contribution to quantum theory with his development of wave mechanics in 1926. Schrödinger's wave equation provided a powerful mathematical framework for describing the behavior of quantum systems. The wave function, derived from Schrödinger's equation, described the probability distribution of finding a particle in a particular state, introducing a probabilistic interpretation to quantum mechanics. Max Born, a German physicist, played a crucial role in interpreting the meaning of the wave function. Born proposed that the square of the wave function represented the probability density of finding a particle in a given location. This interpretation reconciled the mathematical formalism of wave mechanics with the probabilistic nature of quantum phenomena. The combined efforts of these pioneering physicists laid the groundwork for the development of quantum mechanics as a unified and comprehensive theory. In 1926, a landmark conference in Solvay, Belgium, brought together many of the leading minds in physics to discuss the emerging field of quantum mechanics. This gathering included luminaries such as Albert Einstein, Niels Bohr, Max Planck, Werner Heisenberg, and Erwin Schrödinger. The debates and discussions at the Solvay Conference illuminated the challenges and complexities of quantum theory, shaping its future direction. Quantum mechanics introduced a radical departure from classical physics, where particles were no longer deterministic entities with well-defined trajectories. Instead, quantum particles existed as probability distributions described by wave functions, and their properties could only be known probabilistically through measurements. This probabilistic nature gave rise to some perplexing and counterintuitive phenomena. One such phenomenon was quantum entanglement, where two or more particles became correlated in such a way that their properties remained interconnected, even when separated by vast distances. Einstein famously referred to this as "spooky action at a distance" and questioned the completeness of quantum mechanics. Nevertheless, experimental evidence for entanglement continued to mount, challenging classical intuitions and confirming the unique predictions of quantum theory. The development of quantum mechanics also led to a deeper understanding of the behavior of electrons within atoms. Paul Dirac, a British physicist, made significant contributions to quantum mechanics with his formulation of quantum mechanics using the principles of linear algebra. Dirac's mathematical formalism allowed for a more elegant and general description of quantum systems and paved the way for quantum field theory. Quantum field theory, developed by luminaries such as Dirac, Max Planck, and Wolfgang Pauli, extended quantum mechanics to encompass the behavior of fields and particles in a relativistic framework. This framework became essential for describing the behavior of particles at high energies, giving rise to the field of quantum electrodynamics (QED). QED, formulated by Julian Schwinger, Richard Feynman, and Tomonaga Shinichiro, successfully described the electromagnetic interactions of charged particles, and its predictions were in excellent agreement with experimental results. Quantum field theory and QED laid the foundation for the development of the standard model of particle physics, which encompasses the electromagnetic, weak, and strong nuclear forces, as well as the Higgs boson. The development of the standard model represented a triumph of quantum physics, providing a comprehensive framework for understanding the fundamental particles and their interactions. Quantum mechanics and quantum field theory have not only transformed our understanding of the microscopic world but have also yielded numerous technological innovations. Quantum technologies, including lasers, semiconductors, and nuclear magnetic resonance, have become integral parts of modern life and continue to advance our capabilities in communication, computing, and medical imaging. The historical development of quantum physics is a testament to human curiosity, ingenuity, and perseverance in the face of profound scientific challenges. It is a story of paradigm shifts, mathematical elegance, and experimental rigor that has reshaped our view of the universe from the smallest subatomic particles to the vast cosmos. As we delve deeper into the quantum realm, the journey of discovery continues, promising even more profound insights and transformative technologies on the horizon.

Chapter 2: Quantum States and Wavefunctions

Understanding quantum state vectors is at the heart of comprehending the behavior of quantum systems, providing a mathematical framework for describing the quantum state of particles and their properties. Quantum state vectors, also known as wave functions or ket vectors, are essential in quantum mechanics as they represent the complete information about a quantum system. These vectors capture not only the particle's position but also its momentum, energy, spin, and other observable properties. In essence, a quantum state vector encapsulates all possible outcomes of measurements on a quantum system. The mathematical representation of a quantum state vector is often denoted by the symbol $|\psi\rangle$, where ψ represents the state itself. This notation emphasizes that the state vector is a column vector in a complex vector space. Quantum state vectors are elements of a Hilbert space, a mathematical space that accommodates complex vectors with special properties suitable for quantum mechanics. In this space, each possible quantum state corresponds to a unique vector, and the principles of superposition and linearity apply. The principle of superposition states that if $|\psi 1\rangle$ and $|\psi 2\rangle$ are two valid quantum states, then any linear combination of them, such as $a|\psi 1\rangle + b|\psi 2\rangle$, where a and b are complex numbers, is also a valid quantum state. This mathematical concept lies at the core of quantum mechanics, allowing quantum systems to exist in *multiple states simultaneously. The probability of measuring a particular outcome from a quantum state is determined by the square of the magnitude of the coefficient in front of that state in the superposition. The complex nature of these coefficients introduces a unique feature of quantum mechanics: interference, where different paths or states can combine constructively or destructively, affecting measurement outcomes. Quantum state vectors evolve over time according to the Schrödinger equation, a fundamental equation in quantum mechanics. The Schrödinger equation describes how the state vector changes with time and is determined by the Hamiltonian operator, which represents the total energy of the quantum system. Solving the Schrödinger equation allows us to predict how a quantum state will evolve and how its properties will change over time. Quantum state vectors can represent discrete states, such as the energy levels of an electron in an atom, or continuous states, such as the position and momentum of a particle in free space. In the case of discrete states, the quantum state vector is a column vector with a finite number of components, each corresponding to a particular energy level or quantum state. For continuous states, the quantum state vector becomes a mathematical function defined over a continuous range of positions or momenta. This continuous representation is often encountered when dealing with particles in free motion or quantum fields. Quantum state vectors are normalized to ensure that the total probability of all possible outcomes adds up to one. Normalization requires that the inner product of* the state vector with itself, denoted as $\langle\psi|\psi\rangle$, equals one. This normalization condition ensures that the probabilities of all possible measurement outcomes sum to unity, a fundamental requirement in quantum mechanics. When working with normalized quantum state vectors, the coefficients in the superposition represent the probability amplitudes of measuring the corresponding states. The phase of these coefficients plays a crucial role in interference phenomena, as it determines the constructive or destructive interference of different states. Measurement in quantum mechanics plays a pivotal role in collapsing a quantum state from a superposition of possibilities into a single definite outcome. When a measurement is performed on a quantum system, the state vector collapses into one of the possible eigenstates of the measured observable, with the probability of each outcome determined by the square of the coefficient in the superposition. This collapse of the wave function is one of the most enigmatic and debated aspects of quantum mechanics. The act of measurement itself remains a topic of philosophical and experimental investigation, with interpretations ranging from the Copenhagen interpretation to many-worlds theory. Quantum state vectors can be represented in various coordinate systems or bases. The choice of basis depends on the specific problem or observable of interest. For example, in the position basis, the quantum state vector represents the probability amplitude of finding a particle at each position in space. In the momentum basis, it represents the probability amplitude of measuring a certain momentum value. Transformations between different bases are accomplished using mathematical operations known as basis transformations or change of basis. In quantum mechanics, the inner product of two quantum state vectors plays a crucial role in calculating probabilities and expectation values. The inner product, denoted as $\langle \varphi | \psi \rangle$, measures the overlap or similarity between two quantum states. If the inner product is zero, the two states are orthogonal, meaning they have no similarity, while a nonzero inner product indicates some degree of overlap. Quantum state vectors can be used to calculate expectation values, which represent the average outcome of a measurement of a specific observable on a quantum system. The expectation value of an observable A, denoted as $\langle A \rangle$, is calculated as the inner product of the state vector with the operator representing the observable, $\langle \psi | A | \psi \rangle$. This mathematical operation yields the expected value of the measurement outcome for that observable. The concept of quantum state vectors extends beyond single-particle systems and applies to composite systems, such as quantum entanglement, where the quantum state vector describes the entire system. Quantum entanglement occurs when the state of one particle is dependent on the state of another, even when separated by large distances. The entangled state is described by a composite quantum state vector that cannot be separated into individual states for each particle. The study of quantum state vectors is a fundamental aspect of quantum mechanics, and it provides the foundation for understanding the behavior *of quantum systems, from atoms and molecules to the fundamental particles of the universe. These mathematical representations, with their unique properties of superposition, interference, and measurement collapse, underpin the astonishing phenomena and technological advances that characterize the world of quantum physics.*

Chapter 3: Operators and Observables in Quantum Mechanics

Operators play a fundamental role in quantum physics, serving as mathematical entities that represent physical observables and transformations in the quantum world. They are central to understanding how quantum systems evolve, how measurements are made, and how quantum information is manipulated. Operators are represented by mathematical matrices or differential equations that act on quantum state vectors, allowing physicists to calculate and predict the behavior of quantum systems. In the context of quantum mechanics, operators are often associated with physical observables, such as position, momentum, energy, angular momentum, and spin. Each observable corresponds to a specific operator that acts on the quantum state vector to yield a measurable value when measured. For example, the position operator, denoted as $\hat{X}$, acts on the quantum state vector to provide information about the position of a particle. Similarly, the momentum operator, denoted as $\hat{P}$, reveals information about the particle's momentum. The significance of operators lies in their ability to encode the fundamental properties of physical observables and their relationship to quantum states. Operators are often represented as Hermitian matrices, which are square matrices that are equal to their own complex conjugate transpose. This property ensures that the eigenvalues of Hermitian operators are real numbers, *making them suitable for physical observables that yield real measurement outcomes. The eigenstates of Hermitian operators, called eigenfunctions, represent the possible states in which a physical observable can exist, and the corresponding eigenvalues represent the allowed measurement outcomes. Operators also play a crucial role in defining the evolution of quantum systems over time. The Schrödinger equation, a fundamental equation in quantum mechanics, describes how a quantum state vector evolves with time and is determined by the Hamiltonian operator, denoted as $\hat{H}$, which represents the total energy of the quantum system. The time-dependent Schrödinger equation is given by $\hat{H}|\psi(t)\rangle = i\hbar\, \partial|\psi(t)\rangle/\partial t$, where $\hbar$ is the reduced Planck constant and $\partial/\partial t$ represents the time derivative. This equation provides a framework for calculating how quantum systems change and transition between different states as time progresses. Operators can also be used to calculate the expectation values of physical observables. The expectation value of an observable A, denoted as $\langle A \rangle$, is calculated as the inner product of the quantum state vector with the operator representing the observable, $\langle\psi|\hat{A}|\psi\rangle$. This mathematical operation yields the average value of the measurement outcome for that observable, providing a key insight into the behavior of the quantum system. Operators are not limited to representing physical observables; they are also used to describe transformations and operations in quantum mechanics. For example, unitary operators play a crucial role in quantum circuits and quantum computing, where they represent quantum gates that* manipulate quantum states. Unitary operators preserve the normalization of quantum state vectors and ensure that the probabilities of measurement outcomes remain consistent. Quantum gates, which are composed of unitary operators, perform operations such as quantum entanglement, superposition, and measurement on quantum bits or qubits, the basic building blocks of quantum computing. In addition to physical observables and transformations, operators are used to define the fundamental principles of quantum mechanics, such as the Heisenberg uncertainty principle. This principle states that certain pairs of complementary properties, such as position and momentum, cannot be simultaneously measured with arbitrary precision. Mathematically, the uncertainty principle is represented by the commutation relation between the position and momentum operators, $[\hat{X}, \hat{P}] = i\hbar$, where $[A, B]$ denotes the commutator of operators A and B. This relation establishes a fundamental limit on the precision with which certain pairs of observables can be simultaneously known, highlighting the inherent uncertainty in quantum measurements. Operators also provide a powerful tool for solving quantum mechanical problems. By applying operators to quantum state vectors and using mathematical techniques, physicists can derive solutions to the Schrödinger equation and calculate the allowed energy levels and wave functions of quantum systems. Operators, along with the principles of superposition and linearity, allow physicists to explore and understand complex quantum phenomena, from the behavior of electrons in atoms to the dynamics of *quantum fields. Quantum mechanics relies on the algebraic properties of operators and their ability to transform quantum states, making them indispensable tools for describing and predicting the behavior of quantum systems. The mathematical elegance of operators, combined with their physical significance, forms the foundation of quantum physics and underlies the development of quantum technologies that continue to revolutionize our understanding of the universe and our ability to manipulate it at the quantum level.*

Top of Form

Chapter 4: Quantum Measurement and Uncertainty

Operators are the mathematical machinery that underpins the entire edifice of quantum physics, playing a pivotal role in understanding and describing the behavior of quantum systems. In the realm of quantum mechanics, operators represent physical observables and transformations, serving as the bridge between mathematical formalism and the tangible world of quantum phenomena. These mathematical entities are central to unraveling the intricacies of quantum systems, from the behavior of subatomic particles to the properties of quantum fields. Operators are fundamentally linked to physical observables, which are the properties of quantum systems that can be measured and observed. These observables encompass a wide range of physical quantities, including position, momentum, energy, angular momentum, and spin, among others. Each observable corresponds to a specific operator, which, when applied to a quantum state vector, yields a measurable value upon measurement. For example, the position operator, denoted as $\hat{X}$, acts on a quantum state vector to provide information about the position of a particle, while the momentum operator, denoted as $\hat{P}$, reveals information about the particle's momentum. The power of operators lies in their capacity to encapsulate the essential properties of these physical observables and their relationship to quantum states. Operators are often represented as

Hermitian matrices, a class of square matrices that are equal to their own complex conjugate transpose. This property ensures that the eigenvalues of Hermitian operators are real numbers, making them suitable for physical observables that yield real measurement outcomes. The eigenstates of Hermitian operators, known as eigenfunctions, represent the possible states in which a physical observable can exist, with the corresponding eigenvalues denoting the allowed measurement outcomes. In this manner, operators are indispensable in quantum mechanics for quantifying and characterizing the quantum world. Operators also have a vital role in describing the evolution of quantum systems over time. The Schrödinger equation, one of the fundamental equations in quantum mechanics, governs the time evolution of quantum state vectors. This equation articulates how a quantum state vector changes over time and is determined by the Hamiltonian operator, denoted as $\hat{H}$, which represents the total energy of the quantum system. The time-dependent Schrödinger equation is expressed as $\hat{H}|\psi(t)\rangle = i\hbar\, \partial|\psi(t)\rangle/\partial t$, where $\hbar$ represents the reduced Planck constant and $\partial/\partial t$ signifies the time derivative. Solving the Schrödinger equation allows physicists to predict how quantum systems will evolve and transition between different states as time progresses. Operators also feature prominently in calculating the expectation values of physical observables. The expectation value of an observable A, denoted as $\langle A \rangle$, is computed as the inner product of the quantum state vector with the operator corresponding to the observable: $\langle \psi|\hat{A}|\psi\rangle$.

This mathematical operation furnishes the average value of the measurement outcome for that observable, granting valuable insights into the behavior of the quantum system. Operators are not limited to representing physical observables alone; they also describe transformations and operations in quantum mechanics. Unitary operators, for instance, are critical components in quantum circuits and quantum computing. Unitary operators serve as quantum gates, responsible for manipulating quantum states. These operators preserve the normalization of quantum state vectors, ensuring that probabilities of measurement outcomes remain consistent. Quantum gates, composed of unitary operators, perform operations such as quantum entanglement, superposition, and measurement on quantum bits, or qubits, which are the fundamental units of quantum computing. In addition to their roles in describing physical observables and transformations, operators are instrumental in defining the fundamental principles of quantum mechanics, most notably the Heisenberg uncertainty principle. This principle stipulates that certain pairs of complementary properties, such as position and momentum, cannot be simultaneously measured with arbitrary precision. Mathematically, the uncertainty principle is articulated through the commutation relation between the position and momentum operators, $[\hat{X}, \hat{P}] = i\hbar$, where $[A, B]$ denotes the commutator of operators A and B. This relation imposes a fundamental constraint on the precision with which specific pairs of observables can be simultaneously known, highlighting the intrinsic *uncertainty in quantum measurements. Operators also offer a potent toolkit for solving quantum mechanical problems. By applying operators to quantum state vectors and employing mathematical techniques, physicists can derive solutions to the Schrödinger equation. These solutions enable the calculation of allowed energy levels and wave functions for quantum systems. Operators, in conjunction with the principles of superposition and linearity, empower physicists to explore and comprehend complex quantum phenomena, ranging from the behavior of electrons within atoms to the dynamics of quantum fields. In essence, quantum mechanics relies on the algebraic properties of operators and their capacity to transform quantum states, making them indispensable for describing and predicting the behavior of quantum systems. Operators epitomize the mathematical elegance that underpins quantum physics, and their physical significance underscores the development of quantum technologies, which continue to reshape our understanding of the universe and revolutionize our ability to manipulate the quantum realm.*

In the fascinating world of quantum mechanics, the process of measurement stands as a central enigma, often defying our classical intuitions. Unlike classical physics, where measurements typically reveal precise information about a system, quantum measurements have unique characteristics that lead to intriguing consequences. At the heart of the quantum measurement process lies the concept of wavefunction collapse, a phenomenon that challenges our classical understanding of reality. Wavefunction collapse is a fundamental concept in quantum mechanics that describes how the act of measurement can drastically affect the state of a quantum system. In quantum physics, a quantum system is described by its wavefunction, which encodes all the information about the system's possible states. The wavefunction encompasses a range of possibilities, representing a superposition of different states. This superposition is a core feature of quantum mechanics, allowing quantum systems to exist in multiple states simultaneously. However, when we perform a measurement on a quantum system, something peculiar happens—the system appears to "choose" one of its possible states. This apparent choice is the result of wavefunction collapse. Wavefunction collapse is often described as the sudden reduction of a superposed wavefunction to a single, definite state. The act of measurement forces the quantum system into one of its eigenstates, which corresponds to a specific measurement outcome. For example, consider a quantum particle with a superposed wavefunction representing its position. When measured, the particle will appear at a specific position, seemingly chosen at random from its range of possibilities. This abrupt transition from a superposition of states to a single, well-defined state is at the heart of wavefunction collapse. The process of wavefunction collapse has intrigued and perplexed physicists for decades, as it raises profound questions about the nature of reality and the role of observers in quantum systems. One interpretation of wavefunction collapse is the

Copenhagen interpretation, proposed by Niels Bohr and Werner Heisenberg. This interpretation posits that the act of measurement triggers the collapse, and the quantum system adopts a specific state only when observed. According to this view, the observer plays a fundamental role in determining the outcome of a quantum measurement. However, the Copenhagen interpretation does not provide a complete explanation of the underlying mechanisms of wavefunction collapse. Another interpretation, known as the many-worlds interpretation, offers a different perspective. In the many-worlds interpretation, wavefunction collapse doesn't occur. Instead, every possible outcome of a measurement is realized in a separate branch of the universe. In this view, the quantum system remains in a superposition of states, and the act of measurement merely reveals one of the many existing branches. This interpretation suggests that there are countless parallel universes, each corresponding to a different measurement outcome. The debate between these and other interpretations of quantum mechanics continues to be a subject of active research and philosophical inquiry. The wavefunction collapse process also has practical implications in the realm of quantum technology. Quantum computers, for instance, rely on the principles of superposition and entanglement to perform complex calculations at speeds unattainable by classical computers. However, quantum algorithms require the precise manipulation and preservation of quantum states without triggering wavefunction collapse prematurely. Quantum error correction codes and quantum gates are designed to mitigate the effects of wavefunction collapse and enable quantum computers to perform their tasks effectively. In addition to its implications for quantum technology, wavefunction collapse highlights the philosophical challenges of understanding the quantum world. The apparent role of the observer in determining the outcome of measurements raises questions about the nature of reality and the fundamental principles that govern the universe. It challenges our classical intuitions and forces us to grapple with the fundamental nature of quantum systems. Wavefunction collapse remains one of the intriguing and puzzling aspects of quantum mechanics, continually inspiring scientists and philosophers to explore its implications and seek a deeper understanding of the quantum world. As our exploration of quantum mechanics continues, the phenomenon of wavefunction collapse serves as a fascinating and thought-provoking aspect of this captivating field of study. It reminds us that the quantum realm is full of surprises and challenges our preconceived notions about the nature of reality.

Chapter 5: Quantum Entanglement and Superposition

Quantum entanglement is one of the most perplexing and intriguing phenomena in the realm of quantum physics. It's a phenomenon that defies classical intuitions, challenging our understanding of the fundamental nature of the universe. At its core, entanglement describes a peculiar connection between quantum particles that transcends classical notions of physical interaction. When two particles become entangled, their properties become interconnected in a way that seems to defy the boundaries of space and time. This connection persists, even when the entangled particles are separated by vast distances. The phenomenon was famously described by Albert Einstein as "spooky action at a distance," a term that captures the enigmatic nature of entanglement. Entanglement was first introduced in the context of the EPR paradox, a thought experiment proposed by Einstein, Boris Podolsky, and Nathan Rosen in 1935. The EPR paradox highlighted the unusual consequences of entanglement and posed a significant challenge to the emerging field of quantum mechanics. In their thought experiment, the EPR trio imagined two particles created in a way that their properties, such as position and momentum, were perfectly correlated. If one particle's property was measured, it would instantaneously determine the corresponding property of the other particle, regardless of the distance separating them. This apparent violation of the speed of light as an ultimate speed limit in the universe troubled Einstein, who believed that quantum mechanics might be incomplete. However, subsequent experiments, including those by John Bell in the 1960s, confirmed the predictions of quantum mechanics and established the reality of entanglement. Bell's theorem and the subsequent Bell tests demonstrated that entangled particles do indeed exhibit correlations that cannot be explained by classical physics. This experimental evidence supported the view that quantum entanglement is a real and intrinsic feature of the quantum world. The mathematical formalism of quantum mechanics provides a robust framework for understanding entanglement. In quantum mechanics, the state of a system is described by a wavefunction, a complex mathematical function that encodes all the information about the system's properties. When two particles become entangled, their joint wavefunction becomes inseparable, representing a superposition of all possible states of both particles. This entangled state allows for the instant correlation of properties when measured, as each particle's state remains undefined until a measurement is made. One of the most famous experiments demonstrating quantum entanglement is the Einstein-Podolsky-Rosen-Bohm (EPRB) experiment. In the EPRB experiment, a pair of entangled particles, such as photons, are generated and sent in opposite directions to distant detectors. When the properties of one particle are measured, such as its polarization, the other particle's polarization is instantaneously determined, even if it is light-years away. This instantaneous correlation is a hallmark of quantum entanglement and has been confirmed in numerous experiments. Entanglement has profound implications for our understanding of the quantum world and has been harnessed in various applications. Quantum entanglement forms the basis for quantum cryptography, a secure communication method that relies on the fundamental properties of entangled particles to ensure the confidentiality of transmitted information. Additionally, entanglement is at the heart of quantum computing, where qubits (quantum bits) are manipulated in entangled states to perform complex calculations. Quantum teleportation, a process that allows the transfer of quantum information between distant locations, also relies on the principles of entanglement. Furthermore, entanglement plays a crucial role in quantum entanglement swapping, a phenomenon where the entanglement of one pair of particles can be "swapped" onto another pair of particles, even if they have never interacted directly. This phenomenon has implications for quantum communication and the distribution of entanglement over large distances. While quantum entanglement offers exciting opportunities for technological advancements, it also raises profound questions about the nature of reality. The instantaneous correlation of entangled particles challenges our classical conception of cause and effect and suggests a deep interconnectedness at the quantum level. Einstein himself struggled with the implications of entanglement, famously remarking that "God does not play dice with the universe" in response to the probabilistic nature of quantum mechanics. Despite the challenges it poses to our classical intuitions, quantum entanglement has been experimentally validated and stands as a fundamental aspect of quantum physics. It continues to inspire researchers to explore its mysteries and harness its potential for groundbreaking technologies. As we delve deeper into the quantum world, the phenomenon of entanglement reminds us that our understanding of the universe is far from complete and that the quantum realm is full of surprises yet to be unveiled.

Chapter 6: Quantum Dynamics and the Schrödinger Equation

In the realm of quantum physics, the Schrödinger equation stands as a cornerstone of the theoretical framework. It plays a central role in describing the behavior of quantum systems and is fundamental to our understanding of the quantum world. Erwin Schrödinger, an Austrian physicist, introduced this equation in 1926 as a groundbreaking development in the field of quantum mechanics. The Schrödinger equation provides a mathematical framework for describing the time evolution of quantum states. It offers a means to predict how quantum systems, such as particles and atoms, will evolve over time. At its core, the Schrödinger equation is a differential equation that relates the wavefunction of a quantum system to its energy. The wavefunction, denoted by Ψ, encapsulates all the information about the quantum system, including its position, momentum, and other properties. In simple terms, the Schrödinger equation tells us how the wavefunction changes with time, allowing us to make predictions about the behavior of quantum systems. The equation itself takes different forms depending on the context and the type of quantum system under consideration. For a single non-relativistic particle with no spin, the time-dependent Schrödinger equation is written as follows:

$$\textcolor{}{\blacklozenge}\hbar\partial\Psi\partial\textcolor{}{\blacklozenge}=-\hbar22\textcolor{}{\blacklozenge}\nabla2\Psi+\textcolor{}{\blacklozenge}(\textcolor{}{\blacklozenge},\textcolor{}{\blacklozenge})\Psi i\hbar\partial t\partial\Psi=-2m\hbar2$$

$\nabla^2 \Psi + V(r,t)\Psi$ Here, i represents the imaginary unit, $\hbar\hbar$ is the reduced Planck constant, $\partial\Psi\partial$ $\partial t\partial\Psi$ is the time derivative of the wavefunction, m is the mass of the particle, $\nabla^2\nabla^2$ is the Laplacian operator describing spatial derivatives, r is the position vector, and $(\cdot,\cdot)V(r,t)$ is the potential energy. Solving the Schrödinger equation provides us with the wavefunction, which contains all the information needed to understand the quantum system's behavior. The Schrödinger equation is applicable to a wide range of quantum systems, from simple particles to complex atoms and molecules. It has been successfully used to describe the behavior of electrons in atoms, the vibrations of molecules, and the behavior of particles in electromagnetic fields. One of the remarkable features of the Schrödinger equation is its ability to account for the wave-particle duality of quantum objects. In the quantum world, particles like electrons and photons exhibit both wave-like and particle-like behavior. The wavefunction in the Schrödinger equation captures this duality, allowing us to describe particles as waves of probability. This probabilistic interpretation of the wavefunction is a key aspect of quantum mechanics, highlighting the inherently uncertain nature of quantum systems. The Schrödinger equation also connects to the concept of quantization, which is a fundamental characteristic of quantum physics. Quantization means that certain physical properties, such as energy levels in an atom, can only take on discrete, quantized values. This quantization arises naturally from the solutions to the Schrödinger equation for specific systems. For example, when solving the Schrödinger equation for the hydrogen atom, the allowed energy levels correspond to the discrete spectral lines observed in atomic spectra. The Schrödinger equation's success in explaining these quantized phenomena underscores its significance in quantum physics. The equation's formalism extends beyond the time-dependent Schrödinger equation mentioned earlier. There is also a time-independent version, which is used to find the stationary states and energy levels of a quantum system. The time-independent Schrödinger equation takes the following form: ◆◆=◆◆Hψ=Eψ In this equation, ◆H represents the Hamiltonian operator, which is an operator that corresponds to the total energy of the quantum system. ◆ψ is the wavefunction of the system, and ◆E represents the energy of the system in a particular state. By solving this equation, physicists can determine the allowed energy levels and corresponding wavefunctions for a given quantum system. The Schrödinger equation has been incredibly successful in describing a wide range of physical phenomena. It has played a vital role in understanding the behavior of atoms, molecules, and subatomic particles. In the realm of chemistry, the Schrödinger equation has been indispensable for predicting molecular structures and chemical reactions. In quantum mechanics, it has been applied to study the behavior of particles in electromagnetic fields, the formation of energy bands in solid-state physics, and the behavior of particles in quantum wells and quantum dots. Moreover, it provides the foundation for quantum field theory, which describes the behavior of elementary particles and their interactions. Quantum field theory extends the concepts of the Schrödinger equation to encompass the principles of special relativity and incorporates the quantum field concept. While the Schrödinger equation has been immensely successful in explaining many aspects of the quantum world, it is important to note that it is a non-relativistic equation. For particles moving at relativistic speeds, such as those encountered in high-energy particle physics, a more comprehensive framework, such as quantum field theory, is necessary. Despite this limitation, the Schrödinger equation remains a fundamental tool in quantum physics and continues to be a cornerstone of quantum theory. Its elegant mathematical formulation has guided generations of physicists in their quest to understand the intricate behavior of the quantum universe. The Schrödinger equation stands as a testament to the power of mathematics in describing the mysteries of the quantum realm and remains a symbol of the enduring quest to unveil the secrets of the universe at its most fundamental level.

Chapter 7: Advanced Topics in Quantum Mechanics

In the early 20th century, a profound revolution was underway in the world of physics. Albert Einstein had introduced his theory of special relativity in 1905, challenging fundamental notions about space, time, and the nature of the universe. At the same time, quantum mechanics was emerging as a revolutionary theory to describe the behavior of particles at the atomic and subatomic scale.

Both special relativity and quantum mechanics had made significant strides in explaining the physical world, but there was a fundamental problem: they seemed to be incompatible with each other. Special relativity, which provided an elegant framework for understanding the behavior of objects moving at relativistic speeds, relied on a four-dimensional spacetime continuum. In this theory, space and time were intertwined, and the laws of physics were required to be consistent for all observers, regardless of their relative motion. On the other hand, quantum mechanics introduced a probabilistic and inherently uncertain description of particles, where wavefunctions governed their behavior. While quantum mechanics had successfully explained the behavior of particles at the atomic and subatomic scale, it operated within a non-relativistic framework. The equations of non-relativistic quantum mechanics, including the Schrödinger equation, did not account for the principles of special relativity. This discrepancy between the two theories presented a significant challenge to physicists, as it became clear that both theories were essential for understanding the physical world. The need for a consistent framework that could incorporate both the principles of quantum mechanics and special relativity led to the development of relativistic quantum mechanics. Relativistic quantum mechanics, often referred to as relativistic quantum field theory, is a theoretical framework that combines the principles of quantum mechanics with those of special relativity.

It emerged as a solution to the problems posed by the incompatibility of these two groundbreaking theories. One of the key issues that relativistic quantum mechanics addresses is the behavior of particles at high energies and speeds. In classical mechanics, the velocity of an object is limited by the speed of light, c, in a vacuum. However, special relativity showed that as an object approaches the speed of light, its relativistic mass increases, and it requires more and more energy to continue accelerating. This posed a challenge for non-relativistic quantum mechanics, as it could not accurately describe particles, such as electrons, traveling at speeds close to the speed of light. Relativistic quantum mechanics introduced a new theoretical framework that could handle these high-energy scenarios. One of the most significant developments in relativistic quantum mechanics was the introduction of quantum field theory. In this framework, particles are no *longer treated as individual entities with well-defined trajectories, as they are in non-relativistic quantum mechanics. Instead, particles are described as excitations of quantum fields that permeate all of spacetime.*

Each type of particle is associated with a specific quantum field, and interactions between particles are mediated by the exchange of other particles, known as force carriers. Quantum field theory successfully reconciled the principles of quantum mechanics and special relativity by incorporating wavefunctions for particles into the fields themselves. The equations of quantum field theory are relativistically invariant, meaning they remain consistent for all observers, regardless of their relative motion. Quantum field theory also provided a unified framework for describing both particles and fields, making it possible to treat particles and their interactions consistently at all energy levels. One of the most well-known examples of relativistic quantum field theory is quantum electrodynamics (QED), which describes the behavior of electrons, positrons, and photons in electromagnetic interactions. QED successfully explains phenomena such as the Lamb shift and the anomalous magnetic moment of the electron, with remarkable agreement between theory and experiment.

The development of QED and other relativistic quantum field theories marked a significant milestone in the field of theoretical physics. Relativistic quantum mechanics has become the cornerstone of the Standard Model of particle physics, which describes the fundamental particles and their interactions. It has played a crucial role in understanding the behavior of particles in high-energy accelerators, such as those at CERN and Fermilab. Theoretical predictions based on relativistic quantum field theory have been confirmed with remarkable precision in experiments conducted at these facilities.

Furthermore, relativistic quantum mechanics has provided essential insights into the behavior of particles in extreme conditions, such as those found in the early universe or in the vicinity of black holes. It has also been instrumental in explaining the behavior of particles in nuclear reactions and in the study of high-energy astrophysical phenomena.

The development of relativistic quantum mechanics was a significant achievement in the history of physics, as it reconciled two seemingly incompatible theories and provided a comprehensive framework for understanding the behavior of particles and fields in the universe. It demonstrated the power of theoretical physics to develop elegant and consistent descriptions of the natural world, even in the face of apparent contradictions. Relativistic quantum mechanics has not only deepened our understanding of the fundamental forces and particles that make up the universe but has also paved the way for groundbreaking technological advancements, such as particle accelerators and *detectors. In the quest to unravel the mysteries of the universe, relativistic quantum mechanics remains an indispensable tool, guiding physicists toward a more complete understanding of the fundamental laws that govern the cosmos.*

Chapter 8: Quantum Information and Quantum Computing

Quantum information theory stands at the intersection of quantum mechanics and information theory, offering a unique and powerful framework for understanding the transmission and processing of information at the quantum level. This field has garnered significant attention and interest due to its potential to revolutionize the way we communicate, compute, and secure information. At its core, quantum information theory explores how quantum systems can encode, transmit, and process information in ways that classical systems cannot replicate. To appreciate the significance of quantum information theory, it's essential to grasp the fundamental differences between classical and quantum information. In classical information theory, information is typically represented as bits, which can exist in one of two states: 0 or 1. This binary nature forms the foundation of classical digital information processing, where bits are manipulated using logical operations to perform computations and store data. However, quantum systems introduce a paradigm shift by allowing information to be encoded in quantum bits, or qubits. Unlike classical bits, qubits can exist in a superposition of states, meaning they can represent both 0 and 1 simultaneously. This inherent superposition property gives qubits a unique advantage in information processing tasks, as they can perform multiple calculations at once. Moreover, qubits can become entangled, a phenomenon where the state of one qubit becomes dependent on the state of another, even when separated by vast distances. Entanglement allows for the creation of highly correlated quantum states, enabling secure communication and enhanced computational capabilities. One of the most well-known applications of quantum information theory is quantum cryptography, which aims to provide unbreakable encryption methods based on the principles of quantum mechanics. Quantum key distribution (QKD) is a prime example, where entangled qubits are used to generate cryptographic keys that are inherently secure against eavesdropping attempts. The security of QKD relies on the fundamental principle that measuring an entangled qubit disrupts its state, making any unauthorized interception detectable. This guarantees the confidentiality of transmitted information, even in the presence of sophisticated adversaries. Quantum information theory also plays a pivotal role in quantum computing, a field that promises to revolutionize computational capabilities by leveraging the quantum properties of qubits. Quantum computers have the potential to solve complex problems, such as factoring large numbers and simulating quantum systems, much faster than classical computers. This is due to their ability to perform quantum parallelism, where multiple calculations are executed simultaneously through qubit superposition. Furthermore, quantum algorithms, such as Shor's algorithm and Grover's algorithm, have demonstrated significant advantages over their classical *counterparts in specific tasks. While quantum computing is still in its infancy, it holds tremendous promise for a wide range of applications, including optimization problems, drug discovery, and cryptography. Quantum information theory extends beyond cryptography and computing to quantum communication, where qubits are used to transmit information with unprecedented security and efficiency. Quantum teleportation is a remarkable phenomenon enabled by entanglement, allowing the transmission of quantum states from one location to another without physically moving the particles. This has profound implications for future communication networks, where quantum entanglement can be harnessed to establish secure and instantaneous connections. Another application of quantum information theory is quantum error correction, a vital aspect of building practical and reliable quantum computers. Due to the delicate nature of quantum states, errors caused by decoherence and noise can easily disrupt quantum computations. Quantum error correction codes, such as the surface code, provide a means to detect and correct errors in qubit states, making fault-tolerant quantum computing a reality. Quantum information theory also delves into quantum teleportation is a remarkable phenomenon enabled by entanglement, allowing the transmission of quantum states from one location to another without physically moving the particles. This has profound implications for future communication networks, where quantum entanglement can be harnessed to establish secure and instantaneous connections. Another application of quantum information theory is quantum error correction, a vital aspect of building practical and reliable quantum computers. Due to the delicate nature of quantum states, errors caused by decoherence and noise can easily disrupt quantum computations. Quantum error correction codes, such as the surface code, provide a means to detect and correct errors in qubit states, making fault-tolerant quantum computing a reality. Quantum information theory also delves into quantum teleportation is a remarkable phenomenon enabled by entanglement, allowing the transmission of quantum states from one location to another without physically moving the particles. This has profound implications for future communication networks, where quantum entanglement can be harnessed to establish secure and instantaneous connections. Another application of quantum information theory is quantum error correction, a vital aspect of building practical and reliable quantum computers. Due to the delicate nature of quantum states, errors caused by decoherence and noise can easily disrupt quantum computations. Quantum error correction codes, such as the surface code, provide a means to detect and correct errors in qubit states, making fault-tolerant quantum computing a reality. Quantum information theory also delves into quantum teleportation is a remarkable phenomenon enabled by entanglement, allowing the transmission of quantum states from one location to another without physically moving the particles. This has profound implications for future communication networks, where quantum entanglement can be harnessed to establish secure and instantaneous connections. Another application of quantum information theory is quantum error correction, a vital aspect of building practical and reliable quantum computers. Due to the delicate nature of quantum states, errors caused by decoherence and noise can easily disrupt quantum computations. Quantum error correction codes, such as the surface code, provide a means to detect and correct errors in qubit states, making fault-tolerant quantum computing a reality. Quantum information theory also delves into the intriguing concept of quantum teleportation, where the quantum state of one particle can be transferred to another, distant particle through entanglement. This process doesn't involve physical particles being transported but rather the transmission of quantum information that can be reconstructed at the receiving end. Quantum teleportation holds great potential for secure communication and quantum computing. As quantum information theory continues to advance, it opens up new possibilities and challenges in the fields of physics, computer science, and cryptography. It not only expands our understanding of the fundamental principles of the quantum world but also offers practical applications that have the potential to reshape technology and communication in the 21st century. The study of quantum information theory is an exciting journey into the heart of quantum mechanics, where the strange and counterintuitive properties of the quantum realm become the building blocks of a new era in information science.*

Chapter 9: Quantum Field Theory and Relativistic Quantum Mechanics

Quantum Field Theory (QFT) is a fundamental framework in theoretical physics that combines quantum mechanics and special relativity to describe the behavior of particles and fields. It represents a significant advancement from earlier classical field theories, such as classical electromagnetism and classical mechanics. QFT is a powerful tool for understanding the behavior of particles at both the microscopic and macroscopic levels, providing insights into the fundamental forces of the universe. One of the key principles of QFT is the idea of quantization, which means that physical quantities, such as energy and momentum, can only take on discrete, quantized values rather than continuous ones. This concept was first introduced by Max Planck in his work on blackbody radiation, where he proposed that energy is quantized into discrete packets called "quanta." Quantization is a fundamental aspect of QFT and is used to describe the behavior of particles and fields in the quantum realm. In classical physics, fields are described using continuous mathematical functions, such as the electric field in classical electromagnetism. However, in QFT, fields are quantized, meaning that they are described in terms of discrete, quantized excitations known as particles. These particles can be thought of as the smallest units of a field, and they carry specific properties, such as mass, charge, and spin. In the context of QFT, particles are excitations of their respective fields, and their interactions are mediated by the exchange of other particles known as force carriers. One of the most well-known examples of this is the electromagnetic field, which is associated with the photon as its force carrier. Particles with electric charge interact with the electromagnetic field by exchanging photons, leading to the familiar phenomena of electromagnetic attraction and repulsion. Quantum Field Theory also provides a framework for understanding the behavior of particles in the presence of other fields, such as the gravitational field described by general relativity. In this context, the gravitational field is quantized, and particles with mass interact with it through the exchange of hypothetical particles called gravitons. While gravitons have not been experimentally observed, their existence is predicted by the theory. Quantum Field Theory has been incredibly successful in describing the behavior of particles and fields in the quantum realm, and it has provided the foundation for the Standard Model of particle physics. The Standard Model is a theoretical framework that describes the fundamental particles of the universe and their interactions through the electromagnetic, weak, and strong nuclear forces. It has been experimentally validated to a remarkable degree of precision, confirming the predictions of QFT in various particle physics experiments. The Standard Model includes particles like quarks, which make up protons and neutrons, as well as the Higgs boson, which gives mass to other particles through the Higgs mechanism. QFT also plays a crucial role in understanding the behavior of particles in extreme environments, such as those found in high-energy particle accelerators or the early moments of the universe. In these extreme conditions, particles and fields may behave differently than in everyday scenarios, and QFT provides the theoretical tools to make predictions and interpret experimental results. One of the remarkable aspects of QFT is its ability to accommodate the principles of both quantum mechanics and special relativity. Special relativity, developed by Albert Einstein, introduced the concept that the laws of physics are the same for all observers in uniform motion and that the speed of light is constant for all observers. This theory had a profound impact on our understanding of space and time and required a reevaluation of classical physics. QFT successfully incorporates special relativity by treating particles and fields in a way that is consistent with the theory's principles. For example, in QFT, particles are described by quantum fields that obey relativistic equations. The incorporation of special relativity into QFT also leads to the prediction of phenomena such as time dilation and length contraction at high speeds, which have been experimentally confirmed. Quantum Field Theory is not limited to particle physics; it has applications in various branches of physics, including quantum electrodynamics (QED), quantum chromodynamics (QCD), and quantum gravity. QED, developed by Richard Feynman, Julian Schwinger, and Tomonaga Shinichiro, is a specific QFT that describes the electromagnetic interaction between charged particles. It successfully explains phenomena like the Lamb shift and the anomalous magnetic moment of the electron. QCD, on the other hand, is the QFT that describes the strong nuclear force that binds quarks together to form protons, neutrons, and other hadrons. It has a unique property called asymptotic freedom, which explains why quarks and gluons, the particles that mediate the strong force, appear to behave as free particles at very short distances. Quantum gravity, a field that aims to unify quantum mechanics and general relativity to describe the behavior of gravity at the quantum level, is also an area where QFT is essential. However, quantum gravity remains one of the most challenging and unresolved problems in theoretical physics, and researchers continue to seek a consistent and complete theory of quantum gravity. Despite its remarkable successes, Quantum Field Theory also faces several challenges and open questions. One of the primary challenges is the difficulty of reconciling quantum field theories with the principles of general relativity, especially in the context of gravitational interactions. This has led to the development of various approaches to quantum gravity, such as string theory and loop quantum gravity. Another challenge is the presence of infinities in certain calculations within QFT, which require regularization and renormalization techniques to provide meaningful results. These infinities arise in situations involving the self-interactions of particles, and they have been a subject of extensive study and debate in the field. In summary, Quantum Field Theory is a foundational framework in modern physics that combines the principles of quantum mechanics and special relativity to describe the behavior of particles and fields. It has been remarkably successful in explaining the behavior of particles at both the microscopic and macroscopic scales and has played a crucial role in the development of the Standard Model of particle physics. QFT also has applications beyond particle physics, including quantum electrodynamics, quantum chromodynamics, and quantum gravity. While it has achieved numerous successes, it also faces challenges and open questions, such as reconciling with general relativity and addressing infinities in certain calculations. Despite these challenges, QFT remains a fundamental and powerful tool for understanding the fundamental forces and particles that govern the universe.

Chapter 10: Exploring Quantum Gravity and Cosmology

The quest for a quantum theory of gravity represents one of the most profound and challenging endeavors in the history of physics. The two pillars of modern physics, quantum mechanics, and general relativity, have been enormously successful in their respective domains, but they stand in stark contrast when it comes to describing the behavior of gravity at the quantum level. General relativity, developed by Albert Einstein in the early 20th century, presents a geometric description of gravity as the curvature of spacetime caused by mass and energy.

This theory has been verified through numerous experiments and observations and has provided accurate predictions in a variety of astrophysical and cosmological contexts. However, general relativity is fundamentally a classical theory and does not incorporate the principles of quantum mechanics. Quantum mechanics, on the other hand, is the highly successful framework that describes the behavior of matter and forces at the smallest scales. It has been tested rigorously in countless experiments and has given rise to the Standard Model of particle physics, which accounts for the electromagnetic, weak, and strong nuclear forces, as well as the particles that mediate these interactions. Yet, when quantum mechanics is applied to gravity, it reveals a fundamental incompatibility with general relativity. This incompatibility arises from the fact that general relativity treats spacetime as a smooth, continuous fabric, while quantum mechanics operates with discrete, quantized entities. Attempts to reconcile these two pillars of physics have led to various theoretical approaches, each with its own unique insights and challenges. One of the most prominent approaches is string theory, a theoretical framework that posits the existence of tiny, vibrating strings as the fundamental building blocks of the universe. In string theory, the gravitational force emerges naturally from the interactions of these strings, and the theory is inherently quantum mechanical. String theory offers a potential solution to the problem of quantum gravity and has the added benefit of unifying all known forces of nature into a single, coherent framework.

However, string theory remains a highly complex and mathematically demanding theory, with multiple versions and a lack of experimental confirmation. Another approach to quantum gravity is loop quantum gravity, which takes a different route to reconcile general relativity and quantum mechanics. In loop quantum gravity, spacetime is quantized, and the fabric of the universe is thought to be composed of discrete, indivisible units called "quantum loops." These loops interact in a way that gives rise to the geometric properties of spacetime, effectively incorporating gravity into a quantum framework. Loop quantum gravity has made significant progress in addressing some of the challenges posed by quantum gravity, particularly in the context of cosmological and black hole physics. However, like string theory, it also lacks experimental verification and has its own theoretical complexities. In addition to these two main approaches, there are several other quantum gravity theories and models, such as causal sets, non-commutative geometry, and asymptotically safe gravity. Each of these approaches offers its own perspective on how quantum mechanics and gravity may be reconciled, but none has yet been definitively proven or widely accepted.

The search for experimental evidence or observational clues to confirm or rule out these theories remains an ongoing endeavor in the field of theoretical physics. Quantum gravity also has profound implications for our understanding of the very early universe, where both quantum effects and gravitational interactions were immensely significant. Cosmologists are particularly interested in how quantum gravity might have shaped the universe during its initial moments, leading to the phenomena observed in the cosmic microwave background radiation and the large-scale structure of the cosmos. The study of quantum gravity has inspired a rich interplay between theoretical physics, mathematics, and cosmology, pushing the boundaries of our knowledge and challenging our fundamental understanding of the universe. Moreover, it has led to fruitful discussions about the nature of spacetime, the existence of extra dimensions, and the ultimate fate of the universe. One of the fundamental questions in the quest for quantum gravity is whether spacetime itself is a continuous entity or if it undergoes a fundamental discretization at extremely small scales. This question strikes at the heart of the tension between general relativity's smooth spacetime and quantum mechanics' discrete quanta. Various quantum gravity theories propose different answers to this question, and resolving it is crucial for achieving a unified theory of the fundamental forces. Despite the complexity and challenges of the search for a quantum theory of gravity, physicists are driven by the belief that such a theory could revolutionize our understanding of the cosmos and potentially lead to profound technological advancements.

One of the anticipated consequences of quantum gravity is the unification of all fundamental forces into a single, coherent framework. This grand unification would provide a deeper and more complete understanding of the fundamental laws that govern the universe. Moreover, a successful quantum theory of gravity could provide new insights into the nature of black holes, the behavior of matter and energy at extremely high energies, and the structure of spacetime itself. Practical applications of quantum gravity research are also on the horizon, including the development of advanced technologies that leverage our understanding of the quantum nature of gravity. These technologies could have far-reaching implications, from revolutionizing space travel to enabling the creation of powerful quantum computers. In the pursuit of quantum gravity, scientists and theorists continue to explore the deepest mysteries of the universe, pushing the boundaries of human knowledge and seeking a more complete and unified description of the cosmos. The search for a quantum theory of gravity represents a testament to the enduring curiosity and determination of the scientific community to unravel the fundamental secrets of the universe, no matter how challenging the journey may be. Cosmological insights from quantum gravity models offer a fascinating window into the early universe and the nature of cosmic evolution. These models, which aim to unify quantum mechanics and general relativity, provide unique perspectives on the cosmos that depart from classical cosmology.

One of the central questions in cosmology is the origin and behavior of the universe itself. Classical cosmology, based on general relativity, describes the universe's expansion from an initial singularity known as the Big Bang. However, this description reaches a limit when extrapolated back to the very beginning, where the conditions are extreme and quantum effects become dominant. Quantum gravity models step in to address this challenge by providing a framework in which both gravity and quantum mechanics play essential roles. In these models, the universe's evolution is governed by quantum equations that incorporate the discrete nature of spacetime and quantum fluctuations. One of the most intriguing aspects of quantum gravity cosmology is the concept of a "quantum bounce." In classical cosmology, the universe emerges from a singularity at the Big Bang, but quantum gravity models propose an alternative scenario. According to these models, as the universe contracts towards a high-density state, quantum effects become increasingly important and prevent the singularity from forming. Instead, the universe reaches a minimum size and begins to expand again, creating a cyclic or "bouncing" cosmology. This concept challenges our traditional view of cosmic history and opens new avenues for understanding the universe's birth and evolution. Quantum gravity also provides insights into the behavior of matter and energy at extreme energy scales. During the early moments of the universe, when densities and temperatures were extraordinarily high, quantum gravity effects were paramount.

Understanding these effects is crucial for modeling the behavior of the universe during its infancy. Furthermore, quantum gravity may shed light on the nature of dark matter and dark energy, two enigmatic components that dominate the cosmos but remain poorly understood. Some quantum gravity models propose modifications to the laws of gravity at cosmic scales, which could potentially account for the observed accelerated expansion of the universe attributed to dark energy. Exploring these modified gravity theories is essential for unraveling the mysteries of dark energy and its role in cosmic expansion. In addition to addressing the challenges posed by the early universe and dark energy, quantum gravity models also have implications for the study of black holes. These cosmic enigmas are characterized by immense gravitational forces and extreme conditions, where both general relativity and quantum mechanics are crucial. Quantum gravity theories aim to provide a consistent framework for describing the behavior of matter and spacetime in the vicinity of black holes. Such a framework could resolve long-standing paradoxes, such as the information loss problem, which arises when quantum particles fall into a black hole and seem to disappear without a trace. Quantum gravity models suggest that information may be encoded in subtle correlations between particles emitted by black holes, providing a potential solution to this puzzle. Another intriguing aspect of quantum gravity is its connection to the concept of a holographic universe.

In certain models, the universe's information content is thought to be encoded on a lower-dimensional boundary, much like a hologram. This holographic principle suggests that our three-dimensional universe may be a projection of underlying quantum states on a two-dimensional surface. This profound idea has deep implications for our understanding of space, time, and the fundamental nature of reality. Moreover, quantum gravity may offer insights into the fundamental constituents of the universe. In many models, spacetime is composed of discrete building blocks, which can give rise to novel phenomena at ultra-high energies. These building blocks, whether they are strings, loops, or other structures, may be observable through experiments or cosmic observations, providing a direct link between quantum gravity and empirical science. While quantum *gravity models hold great promise for addressing fundamental questions in cosmology, it's important to note that they remain highly theoretical and face significant challenges. Experimental evidence for quantum gravity effects is currently limited, and testing these models under extreme conditions is a daunting task. Nevertheless, ongoing research in astrophysics, particle physics, and cosmology seeks to probe the frontiers of quantum gravity through high-energy experiments and astrophysical observations. In the coming decades, advancements in technology and theoretical insights may bring us closer to detecting quantum gravity signatures in the cosmos. The quest for a complete theory of quantum gravity continues to captivate the imagination of physicists, cosmologists, and scientists worldwide. It represents a frontier of knowledge where the very fabric of the universe is woven together with the principles of quantum mechanics, challenging our preconceptions and reshaping our understanding of the cosmos.*

Cosmological insights from quantum gravity models offer a glimpse into a universe where the boundaries between classical and quantum physics blur, where the birth of the cosmos is marked by a quantum bounce, and where the mysteries of dark matter, dark energy, and black holes are illuminated by the unifying light of a quantum description of gravity. This ongoing exploration promises to reveal the hidden secrets of the universe and redefine our perception of the cosmos.

Conclusion

In this remarkable journey through the realm of quantum physics, we embarked on a voyage of discovery that spanned from the very basics of quantum mechanics to the most advanced concepts in the field. Our four-book bundle, "Quantum Physics Voyage," has been designed to cater to both beginners and those seeking to master the intricacies of quantum physics. Let us take a moment to reflect on the key insights and knowledge we have gained from each of these books.

In Book 1, "Quantum Physics for Beginners: Exploring the Fundamentals of Quantum Mechanics," we set sail on our voyage by unraveling the fundamental principles that underpin the quantum world. We delved into the wave-particle duality, quantum superposition, and the uncertainty principle. These foundational concepts served as our compass, guiding us through the intriguing landscape of quantum mechanics.

Book 2, "From String Theory to Quantum Computing: A Journey Through Quantum Physics," took us on an exhilarating journey through the fascinating realms of string theory and quantum computing. We learned how string theory offers a potential framework for unifying the laws of physics and ventured into the cutting-edge field of quantum computing, where qubits and quantum algorithms hold the promise of revolutionizing computation.

As our voyage continued, Book 3, "Quantum Physics Demystified: From Novice to Quantum Expert," provided us with the tools and knowledge to transition from novice to quantum enthusiast. We explored advanced topics such as quantum states, operators, and experiments, steadily building our expertise in quantum mechanics.

Finally, in Book 4, "Mastering Quantum Physics: From Basics to Advanced Concepts," we reached the pinnacle of our journey. Armed with a deep understanding of quantum physics, we tackled advanced concepts, including quantum field theory, relativistic quantum mechanics, and quantum gravity. We also explored the intriguing connections between quantum physics and the fascinating world of string theory.

As we conclude our voyage through the world of quantum physics, we have not only acquired a profound understanding of the subject but have also gained an appreciation for the profound impact of quantum physics on our understanding of the universe. Quantum physics has revolutionized technology, inspired new frontiers of scientific exploration, and challenged our very notions of reality.

The journey may have been demanding, but the knowledge gained is invaluable. We now stand as explorers of the quantum realm, ready to apply our insights to unravel the mysteries of the universe and *contribute to the ever-evolving landscape of quantum science.*

We hope that the "Quantum Physics Voyage" has served as a valuable resource for both beginners and seasoned enthusiasts, equipping you with the knowledge and curiosity to continue your exploration of the quantum world. As we bid farewell to this voyage, we encourage you to keep your curiosity alive, for the quantum universe holds countless wonders yet to be uncovered.